ADDITIONAL PRAISE FOR *CLIMATE COURAGE*

Climate Courage reframes the narrative on climate change from doom and gloom to opportunity, with a breadth of practical solutions that can appeal to Americans across the political spectrum. This is an excellent summary of how the politics on climate change have evolved and the incredible progress we're making, highlighting coalitions and initiatives that can inspire others. You can't read this and not come away with some optimism!

—TERRY TAMMINEN,
former secretary of the California Environmental Protection Agency and president of 7th Generation Advisors

We have many of the technical solutions we need to address climate change. Andreas takes on a different question: how do we as a society get to the point where we decide to use them? This book, born of hard experience by a longtime clean energy leader, is an excellent guide to finding our way forward on one of the most important questions we face.

—ADAM BROWNING,
founder and executive director of Vote Solar

Karelas provides an uplifting and thoroughly researched study on the magnitude of efforts being made to confront the climate crisis. *Climate Courage* shows that it takes a village to save the planet, and here are so many different examples for how we can do it.

—SARAH SHANLEY HOPE,
founding executive director of the Solutions Project

Climate Courage is a gem of a book. I have been teaching about climate change for close to three decades. No other book comes close to explaining the challenges we face and, more importantly, the profound opportunities we are being offered in building a clean energy future. Read this book and give it to everyone you know.

—PAUL WAPNER,
professor of global environmental politics,
American University, and author of *Is Wildness Over?*

This inspiring look at how and why solar can and should become a leading source of energy—and the diverse leaders on the frontline of change—is a must-read. Karelas is both ethical and practical, and his deep love for people and the planet suffuses every page.

—REV. SALLY G. BINGHAM,
founder and president emeritus of
Interfaith Power and Light and canon for
the environment for the Diocese of California

Looking for a journey of optimism among today's doom-and-gloom headlines about climate change? This book is your tour guide. Thoughtful, well-researched, and thoroughly engaging, *Climate Courage* offers a real-world road map of hope and inspiration about the future of our planet and ourselves.

—CLINT WILDER,
coauthor of *The Clean Tech Revolution*
and *Clean Tech Nation*

The 2020s are supposed to be the decisive decade on climate, but until we're rolling up our sleeves, doing what Andreas describes in vivid detail, we're all "just talking." Here's a book on how to summon up the courage to do, not tell. Read it and you will get the gumption that you need to bridge the divides in America and do the work we need to save ourselves from our own worst impulses. Shine on!

—DANNY KENNEDY,
cofounder of Sungevity and CEO of New Energy Nexus

We now have the ability to harness clean energy technologies as tools for social justice. This book details how the community-based climate movement is making the clean energy transition a reality and tells the stories of the incredible people and organizations that show up for environmental, social, and economic justice every day.

—ERICA MACKIE,
cofounder and CEO of GRID Alternatives

To fight climate change, we need to broaden the conversation and include groups that have been consistently left out, especially those that will be worst affected, including communities of color and low-income communities. *Climate Courage* tells the stories of those of us working to make climate an inclusive movement.

—THE REV. DR. AMBROSE CARROLL,
founder and CEO, Green the Church

In *Climate Courage*, Karelas highlights the human side of the fight against climate change—taking this big, intimidating issue and connecting it to individuals and communities. Tackling the climate crisis is about more than measuring greenhouse gas emissions and creating new technology solutions. Ultimately, it's about people and how they choose to act. *Climate Courage* reminds us that the planet doesn't have a political affiliation; there is just one Earth for us all to protect and enjoy.

—JULIA PYPER,
host and producer of the *Political Climate*
podcast, contributing editor at Greentech Media

CLIMATE COURAGE

CLIMATE COURAGE

HOW TACKLING CLIMATE CHANGE CAN BUILD COMMUNITY, TRANSFORM THE ECONOMY, AND BRIDGE THE POLITICAL DIVIDE IN AMERICA

ANDREAS KARELAS

Foreword by Katharine Hayhoe

BEACON PRESS
BOSTON

BEACON PRESS
Boston, Massachusetts
www.beacon.org

Beacon Press books
are published under the auspices of
the Unitarian Universalist Association of Congregations.

Printed in the United States of America

23 22 21 20 8 7 6 5 4 3 2 1

This book is printed on acid-free paper that meets the uncoated paper
ANSI/NISO specifications for permanence as revised in 1992.

Text design and composition by Kim Arney

Library of Congress Cataloging-in-Publication Data

Names: Karelas, Andreas D., author.
Title: Climate courage : how tackling climate change can
build community, transform the economy, and bridge
the political divide in America / Andreas D. Karelas.
Description: Boston : Beacon Press, [2020] | Includes
bibliographical references and index.
Identifiers: LCCN 2020002381 (print) | LCCN 2020002382 (ebook) |
ISBN 9780807084885 (trade paperback) | ISBN 9780807084892 (ebook)
Subjects: LCSH: Climatic changes—Social aspects—United States. | Climate
change mitigation—Social aspects—United States. | Climate change
mitigation—Political aspects—United States.
Classification: LCC QC903.2.U6 K37 2020 (print) | LCC QC903.2.U6 (ebook) |
DDC 363.738/740973—dc23
LC record available at https://lccn.loc.gov/2020002381
LC ebook record available at https://lccn.loc.gov/2020002382

I dedicate this book to my mom, dad, and brother—
my dear family who taught me the importance
of community, faith, service, caring for the Earth,
and sharing love unconditionally.

CONTENTS

FOREWORD

Climate change. What's the first thing you think of when you hear these words?

For a long time, the archetype of global warming was a polar bear on a piece of melting ice, maybe holding a scrawled sign reading "Halp us Al Gor!" But these days, the first thing that might come to mind is a super-sized hurricane barreling toward the coast of Florida, a record-breaking wildfire scorching thousands of acres—and koalas—in Australia, or a killer heat wave wilting crops across Africa.

Increasingly, climate change is amplifying our extreme weather events in ways that we can see and experience personally. It's making many events bigger, stronger, and a lot more dangerous than they would have been fifty or a hundred years ago. You don't have to go to the Arctic any more to see the evidence of climate change with your own eyes. Its impacts are present wherever we live today.

The changes we're seeing are scary too. As far back as we can go—and using paleoclimate records, we're able to look back millions of years—we've never had this much carbon going into the atmosphere this quickly. The truth is, we are conducting an unprecedented experiment with our planet, the only home we have.

Here's the thing, though. When we survey people across North America, the vast majority agree that climate is changing. The majority even agrees that humans are responsible. And the number of people who are concerned or alarmed about climate change is growing every year. If that's really the case, we have to wonder, then, Why aren't we seeing the meaningful policies and actions we need to solve climate change being implemented everywhere today?

To figure out the answer, consider the polling response to a very different question: "Will global warming harm *you personally?*" Regardless of the growing numbers who acknowledge that the problem is real, the answer to this question is still largely "no." The vast majority of us—even many who are solidly on board with the science and who are enraged by how often it's being neglected by decision makers and politicians—don't think it matters to us personally. Instead, we view climate change as a distant issue: one that matters to polar bears, or to future generations, or to those who live far away from us, or to people who have different values than we do, prizing trees and baby seals over the safety of their families or the health of the economy.

That misperception, which we call *psychological distance,* is only half of the problem. The other half is *solution aversion.* We are told, regularly, that most of the solutions to climate change are harmful to us. Solving climate change, we're told, is too expensive or requires giving up our way of life, or destroying the economy, or an even bigger government than we already have.

Because of this, we remain mired in a vicious and increasingly perilous cycle. Carbon emissions continue to increase. Year after year, we break global temperature records. Scientists publish another report warning of the coming cataclysm. Decision makers and politicians claim these warnings are overblown and repeat the same myths they've used to successfully delay climate action for so many decades: it doesn't matter to us, and it costs too much to fix it.

So how do we break this vicious cycle? We often assume that it's a lack of understanding, rather than psychological distance and solution aversion, that's at the root of the problem. "Everyone just needs to understand the science better, especially the really bad parts," we think. "If they did, they'd realize how bad it really is and how urgently we need to act!"

It's no surprise, then, that many books, movies, and websites about climate change are full of doomsday messaging. The science offers plenty of legitimately frightening examples to pull from. But as Andreas Karelas shares in this book, despair cannot power the long-term, sustained action we need to address climate change and build a better world. Instead, it overwhelms us with fear and anxiety.

Fear may generate a strong, short-term response, but, ultimately, it's more likely to paralyze us. "We do not accept climate change because we wish to avoid the anxiety it generates," George Marshall writes in his book *Don't Even Think About It: Why Our Brains Are Wired to Ignore Climate Change.* And anxiety has a shelf life too. People who are anxious and afraid cannot maintain that state forever. Eventually, they overload and check out.

That's why, to break through this stalemate we're in, we need courage and hope, not discouragement and fear. Where does that hope and courage come from? Not from the science, where every new study seems to show how the climate is changing faster or that its impacts are more widespread than we thought. And they don't often come from politicians, either. Years of failed attempts, broken promises, and rollbacks have left US federal policy worse off today than a decade ago. Neither do they come from a generic, wish-you-well aspiration for a better future that may occur in some mysterious way through no actions of our own.

No, we need specific, actionable solutions. We want to know, What can I do? Is anyone else working on this currently? Can we succeed in stopping climate change and build a greener, more equitable world in the process, or are we truly doomed, as we've heard so many times before?

The answers to these questions come from *people acting:* From people who are doing things, who are sharing things, who are making things happen. From ourselves; from others we may know; from people we don't know who may live on the other side of the world from us but whose story we've learned and who inspire us. From many of the amazing people and organizations and companies and movements that are described, in fact, in this very book.

And that's what this book is all about. It's full of hopeful, inspiring stories of people who are working toward innovative solutions that we can share with everyone we know. Stories like the ones I look for and try to share nearly every day, including with my own faith community.

Being an evangelical in the US can be discouraging, to put it mildly. Between 60 and 70 percent reject the reality of a changing climate, as Christian organizations, media, and political leaders regularly generate and share misinformation about global warming.

At the same time, though, engaging with conservative communities has helped me understand what really matters. It's not more climate data that convinces us to care—rather, it's when we connect the dots between climate change and things we already care about, and share stories of good news solutions, that we become open to learning more. And what's exciting is that more and more unexpected allies are coming to the fore, leading initiatives to support clean energy and fight climate change. You'll learn about many of them in this book!

The truth is, even when life is at its most discouraging, if we set out to look for bright spots, we *will* find them. New technology like electric airplanes, record-setting price drops in solar energy and battery storage, cities and companies that have set 100 percent renewable energy goals, the evolving climate justice movement demonstrating what a just transition from fossil fuels to clean energy looks like, and millions of individual and community-based actions by people just like you—these are actions that begin with a simple conversation where we share our hope for the future and the courage and commitment to get us there.

Without a positive vision of the future and the courage to work for it, the worst ills of climate change will come to pass. But by talking about why it matters and what we can do to fix it, and by working together, individually and collectively, we can create and sustain the hope, the courage, and the will to act that will ensure our world, and that of our children, is not doomed after all.

—Katharine Hayhoe, PhD
Director of the Climate Science Center,
Texas Tech University
2020

INTRODUCTION

Until you dig a hole, you plant a tree, you water it, and make it survive, you haven't done a thing. You're just talking.

—WANGARI MAATHAI[1]

MOUNTAINS OF LITERATURE have been written about the problem of climate change. We've all seen the graphs of CO_2 levels in the atmosphere showing the hockey stick curve shooting up to the right, the highest they've been in fifteen million years.[2] We have heard how perilously close we are to a series of irrevocable tipping points. We know that greenhouse gases trap the Earth's heat in our atmosphere, causing the temperatures to rise, which is destabilizing the delicate balance of our ecosystem. Countless reports have shown us that, as the giant ice sheets at the poles melt, sea levels will rise around the world. Less sea ice is available to reflect the sun's rays back to space, and ancient methane trapped in the permafrost is being released, creating vicious cycles that amplify human-caused warming, increasing the risk of crossing tipping points. We have learned that, as our natural environment becomes increasingly chaotic, humans will face more frequent and more intense fires, floods, storms, droughts, heat waves, and cold spells. More species will go extinct. More land will become desert. More places will become uninhabitable. More people will become refugees fleeing from their homes. More diseases will spread. More wars will break out as people fight over critical resources. In short, life on Earth will become progressively more painful as the climate changes from what we know into an erratic, harsh set of conditions that's completely foreign to us.

I don't need to explain all that here, because it's been documented so well, for so long. The science goes back over two hundred years. There's about as much scientific consensus about anthropogenic climate change as there is about the theory of gravity.[2] My guess is, if you picked up this book, you're not looking for a science lesson on the problem. You already know that there's a problem. You're looking for insights into what we can do about it. You're looking for answers to the question: How do we get on with solving it?

This is a book about climate change *solutions*. It's also about helping America find its way through a divided, troubling time. I believe that these challenges are intimately related. And that through the process of coming together—as citizens and as communities—to solve climate change and create a sustainable economy that uplifts us all, we can remember our shared American values. It will require that we dig deep inside ourselves, put aside our differences, and unite behind a common cause. To do so, we're going to have to find within us what I call "climate courage."

While the science behind climate change is clear, the path to quickly transition our economy, culture, and way of doing things toward a shared vision of a sustainable future is not. These questions sit at the heart of what it means to be human. What are the values that guide our collective behavior in a complex situation like this? How do we work with our neighbors to address a problem that affects us all? How do we find the courage to act in the face of the most daunting threat we've ever faced as a species? And how do we do so when what's causing the problem is so deeply woven into the fabric of our society?

In order to coordinate efforts to quickly fix the problem, we must first agree on a course of action. And that requires building trust. Building trust among diverse groups is tricky, but we're going to have to get very familiar with how to do it. Thankfully, we all have familiarity with trust.

Who do you trust in your life? Perhaps it's someone who has shown that they genuinely care for you. They express concern, show interest in your well-being, and have proven dependable time and again. When someone exhibits compassion toward you consistently, you learn over time that they can be trusted.

Conversely, when you are repeatedly scrutinized, criticized, judged, and penalized, the trust breaks down. Fear, anger, blame, and defensiveness cloud the relationship. Over time, this can become a vicious pattern, resulting in animosity.

Unfortunately, this is the pattern that we now see in our politics repeating itself over and over again. We're living through a time of incredible amounts of change. Technological change, cultural change, political change, and of course, climate change. As times change and people become afraid of swirling uncertainty, we often become more tribal. An "us" and "them" mentality becomes more common. Divisive partisanship is ripping us apart and creating a toxic culture of political sparring and posturing at greater levels than we've seen in a very long time.

Ultimately, climate change will not be solved by climate scientists and engineers caculating how many solar panels we need installed. It will be solved by citizens sitting down and talking to each other over a glass of iced tea about how we need to start taking care of each other, trusting each other, and working together to save this place and everything we love.

It's only when we realize that we're all in this together that we can start to make progress toward fixing the problem. Rowing harder won't make much of a difference if half the folks in the boat are rowing in the opposite direction. In order to all row in the same direction, we must rebuild trust and comradery across the political divide. We have to try and understand each other's points of view and honor each other's humanity. We have to swallow our pride and practice forgiveness and learn to let go of past hurts. As Dr. Martin Luther King Jr. said, "Love is the only force capable of transforming an enemy into a friend. . . . By its very nature, hate destroys and tears down; by its very nature, love creates and builds up. Love transforms with redemptive power."[3]

Protecting our planet, in a way that's good for the economy and good for us as individuals and communities, is something all Americans can suport. A clean energy–powered sustainable economy designed to lift people out of poverty and create a more equitable culture will not only protect the planet from disastrous climate change, but it will create jobs, strengthen our national security, make us more resilient, strengthen our community bonds, and even make us healthier and happier in the process. Solving climate change can be the greatest bipartisan effort this country has ever seen.

The question of our time is whether we have the foresight to put aside our differences for a short while, reach across the aisle to assess the problem collectively, analyze our options, determine how we want to solve it, and make great strides toward doing so, before it is too late. As environmental author, activist, and thought leader Bill McKibben points out, "The one thing never to forget about global warming is that it's a timed test."[4] Once that door is closed, there will be little that we can do. "If we don't solve it soon, we will never solve it, because we will pass a series of irrevocable tipping points—and we're clearly now approaching those deadlines," McKibben reminds us.[5]

While it may seem like a stretch, the truth is that the fate of the human race depends to a large degree on whether Americans can work out their issues: Can we heal the wounds between our parties, find common ground in our political philosophies, reconcile our cultural values, and remember our civility in the process? Can we relearn our ability to debate based on facts and evidence and not political spin and media talking points? Can we value the ability to compromise, to engage in bipartisan solutions, to recognize the humanity in those who see the world differently from us?

This book highlights the bright spots in the climate fight. The stories of people finding common ground. The examples of people from all walks of life, of various political stripes, coming together around common sense solutions that we all can get behind. My hope is that these stories point us in the right direction and give us insights into how to build on that momentum.

This book will lay out in clear terms how the country became polarized around climate change and how we can find middle ground, and it will point to the success stories of those unlikely climate heroes who are bridging the political divide and showing us that climate solutions need not be polarizing. There is a path forward that everyone can support. In the pages that follow, we're going to hear from the people who have discovered their climate courage and are creating solutions in their communities. And we'll learn how we can do the same.

CHAPTER 1

REFRAMING THE NARRATIVE

We need courage, not hope,
to face climate change.

—DR. KATE MARVEL[1]

AS SCARY AS CLIMATE CHANGE IS, I want to let you in on a secret that can hopefully reduce your anxiety a little and give us the courage to look at the problem squarely. To many, this may seem like the best-kept secret in the world. To others who've been paying attention to the trends, this may not be news at all. The secret is this: we are rapidly moving toward a zero emission, renewable energy–based economy that will dramatically improve our lives, our communities, and our society. And it's happening faster than most people think.

When thinking about climate change, many of us freeze up like a deer in headlights. Partly this is due to the overwhelming nature of the problem—climate change is absolutely terrifying. And partly it's because many people think they can't do anything about it. They feel a lack of agency, causing them to disengage from the issue entirely.

As a movement, we first need to reframe the narrative. We need to focus on the *opportunity* more than the *challenge*, because when people are bombarded with fear-based messages, they become immobilized. We must point to achievable solutions, real benefits, and simple ways to get started. Otherwise, collectively, we'll never get off the couch, so to speak.

Recently, I ran a half Ironman triathlon, which includes 70.3 miles of swimming, biking, and running. This was my fifth triathlon. I'm not a particularly fast triathlete, but I get a lot of enjoyment from pushing myself to discover my limits and seeing if I can go the distance. On the bike portion of the race, which is fifty-five miles, I was hurting. It was a hot day, and the course seemed like a never-ending series of hills. I could barely sit on my bike seat anymore. At one point, about forty miles in, a woman passed me and shouted, "Only a few miles left and we'll be able to get off the bike!"

"My goodness, she's right!" I thought. Suddenly, I was a new man. The energy rushed to my legs, and I started pedaling faster with a smile on my face. The race became sheer joy.

A few miles up the road, I was riding next to a woman who had some good music playing through a tiny speaker, which I thanked her for. "We're gonna need it," she responded, "'cause we've got a lo-o-o-ng way to go." My heart sank. My motivation drained right out of me. At that moment, I could feel the weight of the rest of the race on my shoulders. I had so much farther to bike and run.

These two interactions showed me the impact our attitude and focus have over our motivation—and our courage. Even when looking at the same set of facts, how we choose to interpret that information makes all the difference between continuing and giving up, success and defeat.

WHO TO BELIEVE?

When it comes to climate change, the soundbites and opinions are loud. Some say climate change is Armageddon. Others say it's a hoax. The Armageddon narrative is not motivating, uplifting, or inspiring, and thus it's not capable of sustaining interest for long. It's certainly easier to deny the reality of climate change and call it a hoax than to sit with the fear of impending doom. And if we do choose to wrestle with the question of what to do about it, we don't have many options for addressing it head-on, which makes it easier to ignore. Denying the problem exists or ignoring it because we believe we're unable to effect change has led us to our current predicament: most people take little to no action.

What if we could create a new narrative? One that reverses this orientation in our minds? How can we paint a positive picture associated with solving climate change that's intrinsically motivating?

The truth is, yes, it's a very serious situation. But we also have a lot of momentum on our side. We've already come a long way, and we can make it to the finish line if we stay positive, collaborate, and make a good-hearted effort.

BRIDGING THE DIVIDE: US VERSUS THEM

When a conservative hears a liberal tell them that the planet is heating up, and that we have to take urgent action, they're not listening to the message. They're sizing up the messenger: "Does this person share my values? Do they respect my opinions?" they may think. "If not, why should I listen to a word they say?" There's often a perceived air of arrogance and a sense of being talked down to that results in the person receiving the information tuning out all the content.

Former South Carolina congressman Bob Inglis, a Republican, whom I'll discuss more in chapter 4, describes it this way: "What happens often when the left is talking to the right about these issues, it seems like it's coming across as, 'We know better than you do. You're a bunch of hicks from the sticks. We're so much smarter than you are. We've got scientists who tell us this and that. We'll design a regulatory system that will fix things, because we can't trust you to make good decisions.' That's one way it comes across—and it's offensive to conservatives."[2]

Combine this with the media personalities and handful of outspoken elected officials who paint climate change as a liberal conspiracy, and it's no wonder there's still a debate in people's minds.

I think it's crucial for people involved in the climate movement to understand this fundamental point. Liberals from Berkeley with painted faces holding posters with wind turbines on them are not going to convince the rest of America that climate change is real or inspire them to act. More likely, it may exacerbate the feeling of exclusion that Americans experience who don't see themselves as sign-carrying environmentalists. This "us versus them" mindset makes it easy for people to make up their mind before a conversation can even begin.

RED SCIENCE, BLUE SCIENCE

Unfortunately for the planet, when the words climate change are mentioned, the most common association is Al Gore. Don't get me wrong: Al Gore is a peerless environmentalist, and the climate movement owes him an infinite debt of gratitude for the tremendous job he did raising awareness of the issue, particularly with his film *An Inconvenient Truth*. However, a lot of people don't feel positively about Al Gore. He's a politician, a Democrat, a rich white guy. While Al Gore did a great job publicizing climate change, by nature of his role in politics, he also helped politicize it. Before Gore became synonymous with climate change, Republican politicians like President George H. W. Bush were promising to take action on climate change (at UN conventions no less), which we'll hear more about in chapter 3. But by the time *An Inconvenient Truth* came out in 2006, and even more so in the years that followed, climate change had become a deeply divided partisan issue. The Democrats claimed global warming as their fight, Republicans called it a hoax, and belief in a basic scientific fact—that greenhouse gases trap heat in the atmosphere, which is something scientists have documented since the 1800s—was now swayed by political affiliation. Of course, the real driver behind the politicization of climate change was not Al Gore but the covert efforts of the fossil fuel industry, wielding money and power to misinform and sow doubt in the minds of Americans, which I'll explore more in chapter 3. When Al Gore became the face of climate change, it just made it easier to promote a partisan narrative that, suffice to say, has cost us years of inaction.

As we know all too well, America is going through a deeply divided political and cultural moment in history. When it comes to the issue of climate change, Americans have been fiercely loyal to the red team's or the blue team's stance. Thankfully, that's beginning to change. The movement is now starting to benefit from messengers who can speak to new audiences in language that addresses their concerns and values. For example, the pope and other religious leaders are calling for climate action based on spiritual teachings; the Department of Defense lists climate change as a top security threat; and the clean energy industry boasts the fastest-growing job sector in the American economy. As these examples show, addressing climate change is something we can all get behind and, indeed, we already are.

Speaking Each Other's Language

A few years ago, I met a man named George from the Midwest who was conservative and vocal about his politics. He was quick to critique others based on their political views, and he was certainly not shy about telling you how he felt. At one point, he mentioned that he kept a garden, as do I, and we started talking about it. He told me all about the fresh tomatoes he grew, and the hearty green beans, and how delicious they were when he picked them straight from his garden. He also told me all about his canning process for preserving his vegetables through the winter. He lit up talking about the joy it brought him to work with his hands and eat the food that he grew himself. I also love to garden and wanted to learn more about canning. I asked him more about it, and we talked at length about our shared passion. While our politics were different, we had a shared appreciation for growing and preserving our own food. The more we talked, the more we realized how much we had in common.

Rather than immediately dismissing someone else's views on climate change, why don't we first try to understand their point of view and background, and just listen to them? Try to understand where they're coming from? Often, people are more willing to listen to the opinions of others when they feel that their own views have been acknowledged. These can be terribly difficult conversations to have, but they are the most important step we can take to solve climate change.

The good news is that the people who are fighting climate change today increasingly include conservatives who are calling for conservative-based approaches to cool the planet and switch our country to renewable energy. This trend is arguably the most promising development we've had so far. Yet many people remain unaware of their growing advocacy.

THE UNPRECEDENTED OPPORTUNITY

The personal and social benefits of tackling climate change are manifest in many different facets of life. If we use clean energy and energy-efficient technologies at home and in our workplaces, eat more plant-based foods, ride our bikes instead of driving, spend more time in nature and with our families and friends instead of shopping or watching TV, participate in the sharing economy instead of accumulating more things, and live more simple, uncluttered lives, we'll not only stop the planet from warming but we will also live happier, healthier, and more fulfilled lives in communities that are more resilient and more prosperous, communities with more job opportunities and more opportunities for meaningful connection.

"Thus, a life with less—less work, less stuff, less clutter—becomes more: more time for friends, family, community, creativity, civic involvement. Less stress brings more fulfillment and joy. Less rushing brings more satisfaction and balance. Less debt brings more serenity. Less is more," Cecile Andrews and Wanda Urbanska write in their anthology on simple living entitled *Less Is More*.[3]

A Vision from the Not-Too-Distant Future

Imagine you're driving home from work. You pull into your garage and plug a charger into your electric vehicle. The power is coming from a battery bank on the wall that's been charging all day from solar panels on your roof that are effortlessly turning sunshine into electricity. You walk into your house, which is perfectly warmed to the temperature you like, through a combination of passive solar design that keeps your home cool in the summer and warm in the winter, geothermal floor heating that brings heat from the ground into your floors, and an electric heat pump that's also powered by the solar panels on your roof. Your extremely efficient dishwasher and washing machine know when to power up, based on when electric rates are cheapest. When they're done, they pipe the dirty water into your water-efficient toilets. When you and your family want to adjust your smart appliances and indoor climate preferences, you can manage it all from your smartphone.

During the day, when you're at work, after your solar panels have fully charged your battery bank, they begin to export electricity to the community's shared power grid. For every kilowatt-hour they generate, you get paid. You're now selling that power directly to your neighbors every single day without any effort. Whatever they don't need goes to battery storage units in your neighborhood that are part of the local microgrid. The microgrid, thanks to its smart technology, can manage the delivery of energy from many inputs—all the homes that produce and consume energy, the local wind and solar farms that generate power, the electric cars plugging in and sharing their power with the grid—all while incentivizing the smart appliances in your homes to use less electricity during peak times. When there's a storm, instead of experiencing a blackout because a transformer fails across town, your solar panels and battery storage unit will keep the lights on and the power flowing in your home. And when you don't need to drive somewhere, it's easy to get around with your community's robust clean energy–powered public transportation system that efficiently moves people to and fro, without the traffic or emissions.

Jobs, Anyone?

Those solar panels, electric vehicles, smart grids, smart home technology, energy efficiency improvements, and public transportation systems will all be sold, installed, and maintained by members of your community gainfully employed in the clean energy economy. The clean energy economy is already creating a tremendous number of safe, secure, well-paid jobs. In fact, the clean energy industry employs triple the number of people employed by the fossil fuel industry, which we'll learn more about in chapter 5.

In addition, the current electricity grid is based on technology that was invented over one hundred years ago, and many parts of the grid across the US have century-old equipment still in place. This ageing electrical infrastructure needs to be replaced. Shouldn't we give it an upgrade in the process? Modernizing the national electric grid with a focus on decentralized, democratized, distributed renewable energy generation will create a tremendous employment opportunity as well as a safer, more resilient grid.

Is this all a pipe dream? Not even close. This is how many Americans are already living. The more people that embrace this clean energy vision, the faster the costs will come down, and the more efficient and ubiquitous the technology will become. Think about this: the number of Americans with TVs jumped from 9 percent to 65 percent in five short years, from 1950 to 1955. That's how quickly technology can spread.[4] As the solar panels and wind turbines pay for themselves over time, energy costs will become so low that we'll have an abundance of incredibly cheap energy available. Think of all the challenges that we'll be able to overcome when we're not constrained by energy.

MOVING ON

Shifting from an agricultural society to an industrial society was a great human experiment. Our ancestors discovered energy-rich fuels buried under the earth and used them to create a seemingly limitless amount of power that has benefitted our lives and our economies. The ushering in of fossil fuels with the steam engine brought industrial manufacturing, advances in agriculture, medicine, infrastructure, and transportation, and it allowed for the development of modern technology, which has in turn yielded many positive social benefits. For a time, it seemed like the results of the experiment were promising, and that this new way of powering and operating society would last forever.

But now we know better. The two-hundred-year experiment of a fossil fuel-powered industrial society can't be sustained. The extraction and use of fossil fuels is destabilizing our climate and destroying our ecosystems, not to mention that there's only a limited supply left. Our cradle-to-grave, disposable, consumptive, throw-away economy, which turns the earth's resources into waste at increasing rates, cannot be sustained.

Now that we know this, we can learn from it and move on. To put it in business terms, we must pivot, and fast. This isn't easy for massive institutions. Government bureaucracies, corporations, and industries that are heavily invested in old technologies and ways of doing things don't adapt very well to change. They tend to stick to what's familiar and what's been profitable for them in the past, ultimately making them vulnerable to replacement by disruptive innovations.

Empowering Citizens and Communities to Be Changemakers

Nimble teams of empowered decision makers, however, can pivot quickly. In today's world, as citizens we often feel like cogs in a machine. Big companies see us as consumers; politicians see us as voters; the media sees us as viewers. Our human spirits are yearning for the opposite. We want to be useful and to make an impact with our actions. We want to create things, we want to make decisions that positively impact our lives and our communities. We want to innovate and build and produce content. We want to be makers and doers, not takers and users. And that's what's so exciting about the moment we're living in.

To solve climate change, we're going to have to remake the world and reimagine how it works, by running countless new experiments. We have the opportunity to redesign *everything*, from how we eat, to how we get around, to how we power our society, to how we plan our cities, to how we build our buildings, to how we interact with each other, to how we make a living. This is going to require *everyone* to be creative doers, makers, and thinkers. The times challenge us all to be visionaries exploring the upper limits of humanity's potential. They challenge us to explore how we can create new ways of living that are more in harmony with nature, with each other, and with ourselves.

While it may sound like a heavy lift, here's the good news. It's already happening. Despite the efforts of many to keep us stuck in the fossil fuel age, the economics of clean energy, energy efficiency, electric vehicles, smart grids, battery storage, organic farming, and the like are disrupting the current state of technology. As the costs of clean energy technology continue to drop and the fossil fuel industry approaches obsolescence, the question is no longer "How do we create a carbon-free economy?" Rather it is "How can we do so in a way that's seamless, equitable, and fast enough?"

CHAPTER 2

THE PSYCHOLOGY OF CLIMATE CHANGE

It is a long journey from the head to the heart, and an even longer journey from the heart to the hands.

—BARTHOLOMEW I,
Ecumenical Patriarch of Constantinople[1]

AS SOMEONE WHO's deeply committed to solving the climate crisis, I'm often curious to see how it's being treated in the news and talked about in the public sphere. Climate change is certainly getting more attention these days, although some of that attention may not be very helpful to our cause. While it's critical to raise awareness about the serious nature of the problem, we have to assess whether the way we frame the issue is working to win people over and mobilize them to take meaningful action. Sometimes I think certain climate change messages have the exact opposite effect. Take these recent news headlines: "Climate Change Is Accelerating, Bringing World 'Dangerously Close' to Irreversible Change."[2] Or "No One Should Want Their Children to Live in This 'Bleak' Future."[3]

Not exactly pick-me-uppers.

If you walk into your local bookstore, you might see these ominous titles: *The Uninhabitable Earth: Life after Warming* and *Losing Earth: A Recent History*. I have great respect for the authors of these books and others like them. These books are impeccably researched, and they are providing important information to readers. I believe that the intent of the authors is ultimately to inspire and mobilize us to act, not scare the bejeezus out of us. But will they have the desired effect? For some, perhaps. But for most people, will framing the issue with such terrifying titles inspire them? Or immobilize them with fear?

With so much to worry about already in daily life, if I'm not already deeply concerned about climate change, will I even bother to pick up a book like this? Which begs another question: Will framing the issue in this way get *new* people to care about climate change, or is it just preaching to the choir? And for the people who do read these books, are they inspired to act as a result?

Best-selling author Andrew Solomon wrote, "*The Uninhabitable Earth* hits you like a comet, with an overflow of insanely lyrical prose about our pending Armageddon."[4] Pending Armageddon? Are you kidding me? Is that supposed to inspire me to pick up a copy? I've got enough to worry about. No thank you.

I was on the BART train recently and saw an elderly man with a big white beard reading the book *The Uninhabitable Earth*. I couldn't help but ask, "What do you think of the book so far?" He replied, "Extraordinary. Even if you are liberal and know about climate change, you realize how uninformed you are." Curious to hear more, I asked how this made him feel. "Hopeless, because we're not gonna stop it," he said. Then he got off the train.

Hopeless. Here's a man who bought a book, is taking the time to deeply study the issue of climate change, and it's making him feel . . . *hopeless*. Not inspired to take action. Not ready to work for climate solutions in his community. Hopeless.

Recently the *New Yorker* published an article by novelist Jonathan Franzen entitled "What If We Stopped Pretending?" with the tagline "The climate apocalypse is coming. To prepare for it, we need to admit that we can't prevent it."[5] Thankfully, top scientists in the field who disagree with him challenged this perspective immediately. Stanford scientist Mark Jacobson, a leader in clean energy research whom we'll hear from throughout the book, tweeted in response to the article: "No, #globalwarming IS a solvable problem and 61 countries have already passed laws to partly get there."[6] NASA climate scientist Kate Marvel, pulling no punches, quickly penned a response published in *Scientific American* entitled "Shut Up, Franzen." In the article she points out, "Climate change is real, and things will get worse. But because we understand the driver of potential doom, it's a choice, not a foregone conclusion."[7]

Now compare the anxiety-inducing book titles I've just introduced with three other recent book titles: *Climate of Hope: How Cities, Businesses, and Citizens Can Save the Planet*; *Drawdown: The Most Comprehensive Plan Ever Proposed to Reverse Global Warming*; and *Being the Change: Live Well and Spark a Climate Revolution*. Which framing is more likely to motivate you? And how about someone new to the climate movement? The perspective that focuses on the latest doomsday predictions that play into the hands of climate deniers calling us climate alarmists? Or the one that focuses on practical steps we can take right now that improve our lives, build on positive momentum, and demonstrate a clear path toward solutions?

OUR BRAINS ON CLIMATE CHANGE

To better understand how we can mobilize people to solve the problem, we have to understand why we have had limited success to date. While industry and politics have played major roles in delaying action, the real problem is how our brains are wired. The eminent psychologist Daniel Goleman, famous for his work on emotional intelligence, explains his thoughts on climate change inaction this way:

> As we've seen, a blind spot in the human brain may contribute to this mess. Our brain's perceptual apparatus has fine-tuning for a range of attention that has paid off in human survival. While we are equipped with razor-sharp focus on smiles and frowns, growls and babies, as we've seen, we have zero neural radar for the threats to the global systems that support human life. They are too macro or micro for us to notice directly. So when we are faced with news of these global threats, our attention circuits tend to shrug.[8]

Relatedly, Nobel Prize–winning psychologist Daniel Kahneman, author of *Thinking, Fast and Slow*, is "deeply pessimistic" about climate change and our psychological capacity to address the problem.[9] In his book *Don't Even Think About It*, climate change communications specialist George Marshall describes his interview with Kahneman about human psychology and climate change (I've italicized certain text for emphasis):

> His concerns are threefold. First, *climate change lacks salience*—by which he means the qualities that mark it as prominent of demanding attention. . . . Kahneman argues that the greatest salience belongs to threats that are concrete, immediate, and indisputable—for instance, a car out of control driving right at you. By contrast, climate change is, he says, abstract, distant, invisible, and disputed.
>
> The second problem, he notes, is that *dealing with climate change requires that people accept certain short-term costs and reductions in their living standards* in order to mitigate against higher but uncertain losses that are far in the future. This is a combination that, he fears, is exceptionally hard for us to accept.
>
> Third, *information about climate change seems uncertain and contested.* As long as that remains the case, he says, "people will score it as a draw, even if there is a National Academy on one side and some cranks on the other."[10]

Kahneman goes on to say, "To mobilize people, *this has to become an emotional issue.* It has to have the immediacy *and* salience. A distant, abstract, and disputed threat just doesn't have the necessary characteristics for seriously mobilizing public opinion."[11] When Marshall asks Kahneman if people could change their behavior if they understood these cognitive biases better, he replied, "No amount of psychological awareness will overcome people's reluctance to lower their standard of living.[12]

Professor Kahneman has put forth some key points about how humans typically behave and has arrived at a bleak conclusion based on the current perception of climate change in the public mind. But I don't think that his predictions are set in stone. We can make this an emotional issue. We can work to reframe climate change to address the concerns he outlines. In other words, we can use his understanding of human psychology to change the current narrative around climate change and thus, I believe, to change our future.

Let's take his arguments one by one.

Salience

For many years, the negative impacts of climate change have seemed far off in the future. However, these negative impacts are far more salient today than they were even last year, and certainly the year before that. Wildfires of epic proportions are ravaging the American West, causing great damage to individual homes and even entire towns. Historic floods are washing away people's homes in Colorado. Deep droughts are drying up crops in the Midwest. Heat waves and cold spells are breaking records for extreme weather year after year. And ever more powerful hurricanes and superstorms continue to slam into our coasts.

In 2019, Hurricane Dorian, a category five hurricane, the fifth to make landfall in the US in the prior three years, devastated the Bahamas. Twenty-five hundred people went missing, seventy thousand people were left homeless, and the storm caused $7 billion in damages.[13] Many climate refugees, whose homes and towns were obliterated, came to the United States seeking refuge and were denied it. In 2019 Deke Arndt, chief of the monitoring branch at NOAA's National Centers for Environmental Information, told the *Washington Post* that, in the previous five years, "we had about twice the number of billion-dollar disasters than we have in an average year over the last 40 years or so."[14]

While I'm highlighting the devastation of climate change–related disasters here, I'm not suggesting that our broader discussions should be focused on it. I am merely trying to demonstrate that climate change is salient to Americans today and increasingly so. Do you agree?

Standard of Living

Kahneman argues that people won't tackle climate change if it results in higher costs or a lower standard of living. I agree with him. But what I'll argue in the remaining pages of this book is that solving climate change will *reduce* our costs and *increase* our standard of living.

Mitigating climate change doesn't have to be punitive. Rather, it will save us money, improve our health, increase our happiness, build stronger bonds in our communities, create jobs, and dramatically elevate our well-being. We need to shift the narrative to highlight the benefits of action instead of the threats of inaction.

Uncertainty

Kahneman believes that uncertainty in the public perception of climate science will result in inaction. I would agree with that assumption as well. For many years, while various sources sowed doubt about climate science in the American mind, we took very little action. But as we'll learn in the coming chapters, certainty about climate science has dramatically *increased* across both parties in just the last few years, in no small part due to the increased impacts of climate change that Americans are seeing in their communities. Among Americans, 73 percent now believe global warming is happening and 62 percent correctly understand that it's caused by humans.[15] While there's certainly still a gap between what the public believes and what the scientists are telling us—due to a large-scale misinformation campaign, which we'll also explore in the coming chapters—we don't need everyone to believe the science. The majority of Americans, if empowered with the proper tools, can create climate solutions in their communities that will save them money, create jobs, increase their security, and build resiliency. Once those trends start hitting the airwaves, you won't need to believe in climate change in order to follow suit. We'll hear about leaders across the country who have gone 100 percent renewable, not to create a climate impact but to create an economic impact. And their neighbors will want to do the same.

In all fairness, changing the public's perception of climate change will be a difficult task. To guide our efforts, let's explore the work of some of today's top thinkers on how to use appropriate messaging to inspire and motivate people to take action.

OUR EMOTIONAL BRAIN AND OUR RATIONAL BRAIN

As the data will demonstrate throughout this book, solving climate change is not a technological challenge. We have all the technology we need to create a sustainable, carbon-free, regenerative economy. It's also not a financial challenge. We have all the money necessary to finance the transition. The real challenge is how to shift the collective will of our nation. To do that we must transform the hearts and minds of Americans.

Transforming hearts and minds is not an easy task, but it is doable. It starts with thoughtful, even artful, messaging. This is where the climate movement has been outmatched by those that wish to maintain the status quo polluting energy economy. As Chip and Dan Heath, business researchers and the authors of *Switch: How to Change Things When Change Is Hard*, point out, "For individuals' behavior to change, you've got to influence not only their environment but their hearts and minds. The problem is this: Often the heart and mind disagree. Fervently."[16]

Leadership expert Simon Sinek, author of *Start with Why*, breaks down the metaphor this way: "The heart represents the limbic, feeling part of the brain, and the mind is the rational, language center. Most . . . are quite adept at winning minds; all that requires is a comparison of all the features and benefits. Winning hearts, however, takes more work."[17]

Consider smoking. Even after the public knew that smoking cigarettes caused cancer and other ill-health effects, it took a very long time for society to reduce its collective smoking habit. Why? Because our behavior is not controlled by our rational decision-making mind—our behavior is controlled by our emotional, more impulsive side. We like to think of ourselves as rational actors, in control of our decisions, using logic to inform our understanding of the world.

The truth is, much of our thinking, behaviors, and decision-making is automatic when our emotions are involved.

To illustrate the difference, psychologists have come to describe the human brain as having two ways of thinking, which can be thought of as our emotional brain and our rational brain. The emotional part of our brain, the limbic system, is more ancient, evolutionarily. It's mainly concerned with the survival of the organism and those patterns of behavior we consider instinct. For example, if a saber-toothed tiger is approaching, you have to act immediately—you don't have the time to think through your options. Your instincts take over and your limbic brain decides whether to run, hide, or fight, popularly known as the "fight or flight" mechanism. The limbic system is also the area of the brain responsible for our emotions and feelings.

According to Dr. Rand Swenson of Dartmouth Medical School, "The limbic system is a convenient way of describing several functionally and anatomically interconnected . . . structures that . . . serve several functions[;] however most have to do with control of functions necessary for self[-]preservation and species preservation."[18] Simon Sinek writes that "the limbic brain is responsible for all our feelings, such as trust and loyalty. It is also responsible for all human behavior and all our decision making."[19]

The part of the human brain that has developed more recently evolutionarily—the neocortex, and specifically the prefrontal cortex—controls language, is capable of rational thought, allows us to think into the future, and is largely what distinguishes humans from the rest of the animal kingdom. It's the area of our brain we use to plan a trip, design a house, or write a poem. Psychologist Piers Steel writes, "The limbic system focuses on the now while the prefrontal cortex deals with longer-term concerns."[20]

While both parts of our brain (our limbic system and prefrontal cortex) are of critical importance, they are often at odds with each other. Different psychologists have different ways of discussing these often conflicting ways of thinking, but they generally agree on the phenomenon. Daniel Kahneman describes the two different modes of thinking as System 1 and System 2. System 1 is our automatic, instinctual limbic brain thinking, and System 2 represents our more rational, logical prefrontal cortex thinking. Daniel Goleman, on the other hand, characterizes our instinctual way of thinking as "bottom-up." Our rational thinking, he refers to as "top-down."

My favorite analogy to describe the relationship between the two portions of the human brain is the Rider and the Elephant, coined by Jonathan Haidt in the 2006 book *The Happiness Hypothesis: Finding Modern Truth in Ancient Wisdom* and further popularized by Chip and Dan Heath in *Switch*. The Elephant represents System 1, the bottom-up, limbic, emotional brain. The Rider represents System 2, the top-down, prefrontal cortex, rational brain. If you think of a person riding an elephant, the person may be holding the reins and doing their best to guide the elephant, but the elephant is really in control of the situation.

The Heath brothers explain Haidt's analogy this way:

> Our emotional side is an Elephant and our rational side is its Rider. Perched atop the Elephant, the Rider holds the reins and seems to be the leader. But the Rider's control is precarious because the Rider is so small relative to the Elephant. Anytime the six-ton Elephant and the Rider disagree about which direction to go, the Rider is going to lose. He's completely overmatched. . . . The weakness of the Elephant, our emotional and instinctive side, is clear: It's lazy and skittish, often looking for the quick payoff (ice cream cone) over the long-term payoff (being thin). . . . The Elephant's hunger for instant gratification is the opposite of the Rider's strength, which is the ability to think long-term, to plan, to think beyond the moment (all the things your pet can't do).[21]

Daniel Kahneman puts it another way, writing that if this concept was made into a movie, "System 2 would be a supporting character who believes herself to be the hero."[22] In other words, our Rider thinks they're the one making things happen, when more often they're just along for the ride.

Both systems of thinking are extremely important and play vital roles in our lives. But here's the big takeaway: our emotional mind makes the decisions. The rational mind is usually left trying to rationalize the behavior the emotional mind has already decided upon. That means that in order to change behavior, whether on an individual level or a societal level, we must appeal to the *emotional* part of our brains *first*. If our emotional brain is scared, overwhelmed, or for whatever reason gets a funny feeling about the change being prescribed, it will bolt in the opposite direction and take the rational mind with it. Only when our emotional mind feels safe and aligned with, and accepting of, the message and the messenger can the rational mind be included in the conversation.

In order to reach people, we have to speak to their emotions, not their logic, say the Heath brothers:

> When people push for change and it doesn't happen, they often chalk it up to a lack of understanding. . . . A scientist says, "If we could just get Congress to *understand* the dangers of global warming, they'd surely take legislative action."
>
> But when people fail to change, it's not usually because of an understanding problem. Smokers understand that cigarettes are unhealthy, but they don't quit. . . .
>
> At some level we understand this tension. We know there's a difference between knowing how to act and being motivated to act. But when it comes time to change the behavior of other people, our first instinct is to teach them something. *Smoking is really unhealthy!* . . . We speak to the Rider when we should be speaking to the Elephant.[23]

We know what we need to do, then. The question is how. Based on the research of thought leaders in this field, I've mapped out a strategy that I think we can adapt to the fight against climate change:

Connect to the emotional brain first: The first thing to do when discussing climate change is to connect with people at an emotional level around shared beliefs. This allows their emotional minds to engage and their rational minds to hear what you're saying.

Stay positive: Focus your message on success stories that inspire. Try not to freak people out. Stay away from doom and gloom and how daunting the challenge is that lies ahead. Instead focus on what's working.

Empower: Be solutions oriented. Give concrete examples of things that people can do that actually make a difference.

Focus on Community: Remind people that they're not alone. We are stronger when we act together. Focus on the shared values and beliefs of the community they identify with.

Now let's explore each of these in a bit more detail.

Connect to the Emotional Brain First

Simon Sinek points out in his book *Start with Why* that what distinguishes great leaders and companies from the rest is their ability to put into words their *why*—to clearly communicate the purpose, the mission, the inspiration behind what they're doing. When the *why* is made clear, the Elephant, our emotional brain, assesses the *why* to see if it agrees. If it does, it will continue listening. If not, you're out of luck. When we start with *why*, he writes, "we're talking directly to the part of the brain that controls decision-making."[24]

The Heath brothers agree: "Because of the uncertainty that change brings, the Elephant is reluctant to move, and analytical arguments will not overcome the reluctance."[25] They remind us to "motivate the Elephant" by engaging people's emotional side. "The challenge is to get the Elephant moving," they write, "even if the movement is slow at first."[26]

Sinek offers a memorable example. He points out that Dr. Martin Luther King Jr. "gave the 'I Have a Dream' speech, not the 'I Have a Plan' speech. It was a statement of purpose and not a comprehensive twelve-point plan to achieving civil rights in America."[27] He was speaking to our emotional minds, not our rational minds. And because of that, he was able to spark a movement. Dr. King's "belief was bigger than the civil rights movement. It was about all of mankind and how we treat each other. . . . People heard his beliefs and his words touched them deep inside. Those who believed what he believed took that cause and made it their own. And they told people what they believed. And those people told others what they believed."[28]

What about solving climate change stirs you emotionally? Can you put it into words? Can you communicate it in a way that might resonate with someone whose politics are different from yours? Something that speaks to what they really care about at an emotional level? Learning how to do that is one of the biggest opportunities we have to create change.

Stay Positive

If you only take one thing away from this book, let it be this: we've got to stay positive, even though the odds are stacked against us.

Picture the following scenario: You're the coach of a high school basketball team. It's the fourth quarter. There are only five minutes left in the game. You're down by ten points, and the other team has been dominating all game. You call a time-out. What do you say to your players? If you follow the tone of many climate change messages, you'd say something like, "The data shows that if we continue playing this way, we're going to lose by a lot. In fact, the chances of us turning this around are slim to none. This is going to be a devastating loss to this team and this community. My prediction is that your friends will stop liking you and you'll be very unpopular. If we project twenty years into the future, I see you all living miserable lives as a result of losing this game."

Or you could say something like, "Yeah, I know we're down, but we've been in tough spots like this before. These next five minutes are about heart. I know how much this team cares. I know that we've got what it takes to rise to the challenge and win this game. Take it one play at a time and keep your eyes on the prize. We've got this!"

If you were a player, which of these messages would you rather hear? Which would inspire you to play better and thus give you a better chance of winning? Daniel Goleman describes this phenomenon of positive versus negative framing as it relates to climate messaging:

> Our impacts on the planet are inherently guilt-inducing and depressing. . . . Focusing on what's wrong about what we do activates circuitry for distressing emotions. Emotions, remember, guide our attention. And attention glides away from the unpleasant.
>
> I used to think that complete transparency about the negative impacts of what we do and buy—knowing our eco-footprints—would in itself create a market force that would encourage us all to vote with our dollars by buying better alternatives. Sounded like a good idea—but I neglected a psychological fact. Negative focus leads to discouragement and disengagement. When our neural centers for distress take over, our focus shifts to the distress itself, and how to ease it. We long to tune out.
>
> So instead we need a positive lens.[29]

We long to tune out. Chip Heath uses a vivid analogy to describe our human tendency to avoid things that distress us: "There's a danger whenever you get a big, scary issue. Say you're a smoker and I tell you about lung cancer. There are two things you can do. You can change—but it turns out empirically the other, more common pattern of behavior is you tune me out. The commercial shows you blackened lungs—well, you can either change your behavior or stop listening to those commercials."[30] This is exactly what happens when we tell people about climate change. While I'm glad that major media outlets are speaking more plainly about the issue, what effect is their coverage having? Seeing a terrifying headline about climate change for the thirtieth time isn't going to suddenly change your mind about the issue. You're already used to tuning it out.

There is another risk associated with negative messaging: you do something trivial to ameliorate climate change to help yourself sleep better at night and then go back to living your life. "Because fear is a negative feeling," Goleman says, "people will take just enough action to change their mood for the better—then ignore it."[31]

Reversing this trend requires messaging focused on climate solutions: a positive frame that emphasizes our successes to date and describes how we build momentum toward larger goals. In his book

The Power of Habit, the Pulitzer Prize–winning reporter Charles Duhigg tells the story of Tom, who had struggled with losing weight. It was when he started looking at the progress he was making that his emotional brain was able to *believe success was possible*: "When I saw those first few pounds disappear, there was this immediate sense of excitement like, wow, I'm really *doing* something. It made it easier to believe this would work, that I could actually succeed at losing weight."[32]

In *Switch*, the authors draw on the work of psychologist Barbara Fredrickson when explaining that, as a survival instinct, "negative emotions tend to have a 'narrowing effect' on our thoughts. . . . Fear and anger and disgust give us a sharp focus—which is the same thing as putting on blinders."[33] When in critical danger, that's a good thing to have, to keep us focused on the problem at hand and not thinking about what's for lunch tomorrow. Positive emotions, on the other hand "are designed to '*broaden* and *build*' our repertoire of thoughts and action. Joy, for example, makes us want to play. Play doesn't have a script, it *broadens* the kinds of things we consider doing. We become willing to fool around, to explore or invent new activities."[34]

Because climate change is such a big problem, in order to address it, we need the human abilities Fredrickson highlights: play, inventiveness, the ability to envision new possibilities. Heath and Heath conclude, "To solve bigger, more ambiguous problems, we need to encourage open minds, creativity, and hope."[35]

Focusing on positive emotions helps keep our audience engaged and inspired. The Heath brothers believe that a "sense of progress is critical, because the Elephant in us is easily demoralized. It's easily spooked, easily derailed, and for that reason, it needs reassurance, even for the very first step of the journey."[36] They advise us to search for "*bright spots*—successful efforts worth emulating."[37] This process, known as "appreciative inquiry," suggests we ask what's working instead of what's not working. Instead of focusing on the latest scientific report that says things are worse than ever, we might tell the story of renewable energy being deployed at breakneck speed around the globe. And the story of the latest school or municipality that's divested from fossil fuels. And the story of the latest city to commit to 100 percent renewable energy. We'll describe the breakthroughs in battery technology that are driving costs down. We'll tell the story of electric vehicles replacing combustion engines around the world. We'll tell the stories about how clean energy jobs are changing people's lives in communities across the country.

Empower

As Lao Tzu wrote in the Tao Te Ching, "A journey of a thousand miles begins with a single step." So it is with solving climate change. While transitioning to an equitable, sustainable economy powered by clean energy will be a massive undertaking, that process is made up of small steps. In order to empower people to get involved, we have to focus on the small but meaningful actions that people can take, cities can take, companies can take, elected officials can take, to get that process started.

One of the biggest dangers we face is debating the perfect plan to solve climate change until we're blue in the face and, in the meantime, not making any progress. Instead, we need to start small and focus on easy-to-achieve targets. With each success, we'll build more momentum, allowing us to set more ambitious goals and allowing the impact of our actions to grow exponentially. As our collective Elephant brain sees the progress being made, it will begin to *believe* that change really is possible. That's when we'll make our biggest strides.

In *Switch*, the Heath brothers describe a village in Vietnam where the children were suffering from chronic malnutrition. After finding a few healthy children among them, a group of doctors studied what they ate and discovered key ingredients in their diet that provided important nutrients that the other children were missing. These ingredients were locally sourced and available to all, so the doctors started a cooking class that taught mothers to use these local recipes to improve their children's health. The mothers were elated as a result of their empowerment: they could make their children healthier, and it wasn't even that hard to do.[38]

Contrast this scenario with the messages we often hear about climate change and what the public comes to believe as a result. It's the opposite, usually along the lines of "Solving climate change will be very hard, and there's nothing I can do about it." Like the Vietnamese mothers, all people need to feel empowered. They need a way to take action that is both simple—something they can realistically see themselves doing—and meaningful. They also need a place to start.

As Sir Isaac Newton's first law says, an object in motion tends to stay in motion. It's getting started, overpowering the inertia, and beginning to move that can be the biggest barrier. Once you're moving, momentum can be a powerful ally. That's why the Heath brothers suggest a strategy they call "shrink the change." Shrinking the change means making the task easily achievable so that we don't fear getting started: "Picture taking a high-jump bar and lowering it so far that it can be stepped over. If you want a reluctant Elephant to get moving, you need to shrink the change."[39]

A great example of this idea involves getting out of debt. When someone is overwhelmed by debt, the initial instinct is to calculate all the debts owed and work hard to pay off the largest debt first and work your way down the list. But paying off the largest debt is not easy, and if we retreat from it, the process stalls. However, if we start with the smallest debt, we can easily see the progress we're making, even in a short time. "If you pay $185 toward a $20,000 debt on a high-interest credit card," Heath and Heath write in *Switch*, "you're still going to feel hopeless. But if you completely pay off $185 overdue utility bill, you can cross it off your list. You've won a victory over debt."[40] The initial change should be "small enough that [you] can't help but score a victory. Once people clean a single room, or pay off a single debt, their dread starts to dissipate, and their progress begins to snowball."[41] By taking small steps, "you're moving forward, and even better, you're getting more confident in your ability to keep moving forward."[42]

Starting small reduces the barriers to getting started. Over time, as we build momentum, taking on larger goals gets easier. With each victory, our emotional brain, the Elephant, becomes more convinced that reaching bigger goals is possible, allowing us to create big changes faster than we might think. By not letting the perfect be the enemy of the good, we understand that our climate action plans, climate policies, and renewable energy targets can evolve and scale over time. But we have to get started. We have to set a clear path toward a sustainable economy, get buy-in from as many Americans as possible, and get moving. If we shrink the change, especially to win over those who are more reluctant, we can take our first steps to decarbonize the economy and see what success looks like. This will alleviate fears and naysaying, and, with each iteration, we'll be able to build on that success with more audacious goals.

Focus on Community

Climate communicator George Marshall, whom we heard from earlier in this chapter, writes, "Even the most unconvinced people can be persuaded by trusted peers who understand their values and can use their common language."[43] Community ties play an important role in any social movement, but when it comes to climate change, the value is even more pronounced. Climate change is a collective problem, caused by collective behavior. Therefore, the only way we can solve climate change is by activating groups of people to assign value to sustainability as part of their community identity. This is critical for two big reasons.

First, community ties let us know that a new behavior or set of values is something that's supported by the group. If we're trying not to freak out our emotional brains, we have to focus our climate action efforts at the community level, so people can see how others behave and thus feel safe participating. The moment someone thinks that if they support climate action, they'll be shunned from the group is the moment they'll tune out.

Second, because it's a scary problem, and change in general can be scary, we want to feel like we're not alone. We want to see that there are other people who are willing to go the distance with us. Thus, it's critical that climate messaging be focused not so much on individuals and what they can do alone, but what groups can do collectively. This is a strategy successfully employed by Alcoholics Anonymous. Charles Duhigg, in his book *The Power of Habit*, writes that "for habits to permanently change, people must believe that change is feasible. The same process that makes AA so effective—the power of a group to teach individuals how to believe—happens whenever people come together to help one another change. Belief is easier when it occurs within a community."[44]

Let's revisit Dr. King and the civil rights movement. We learned earlier that one of the reasons the civil rights movement was so successful was that it connected with people's emotional brain first. It was based on an inspiring vision that resonated with people who shared values and a dream of a more equitable world. Looking around you and seeing that sense of comradery is incredibly empowering and exactly what is needed to create change.

Simon Sinek points out that "Anyone who was drawn to hear Dr. Martin Luther King Jr. give his 'I Have a Dream' speech, regardless of race, religion or sex, stood together in that crowd as brothers and sisters, bonded by their shared values and beliefs. They knew they belonged together because they could feel it in their gut."[45] Sinek goes on to say, "Everyone there that day . . . trusted each other. It was that trust, that common bond, that shared belief that fueled a movement that would change a nation."[46]

"At the root of many movements," Charles Duhigg says, "is a three-part process that historians and sociologists say shows up again and again" [bullets and italics are mine]:

- A movement *starts* because of the social habits of friendship and the strong ties between close acquaintances.
- It *grows* because of the habits of a community, and the weak ties that hold neighborhoods and clans together.
- And it *endures* because a movement's leaders give participants new habits that create a fresh sense of identity and a feeling of ownership.[47]

This third point is critical. The reason new habits continue to spread is because participants feel empowered. They feel a sense of ownership of the movement. And this prompts them to incorporate the new habits into their sense of personal identity. Others who are part of that group can then emulate the behavior, incorporate the new behaviors into their own identity, and continue the process. For example, if your friends start gardening, you might be more inclined to give it a shot yourself, since it fits in with the identity of your community, and eventually being a gardener becomes a part of your own identity.

Regarding the Montgomery bus boycott, Duhigg argues, "Embedded within King's philosophy was a set of new behaviors that converted participants from followers into self-directing leaders."[47] As King gave protesters "a new sense of self-identity, the protest became a movement fueled by people who were acting because they had taken ownership of a historic event. And that social pattern, over time, became automatic and expanded to other places and groups of students and protesters whom King never met, but who could take on leadership of the movement simply by watching how its participants habitually behaved."[49]

At our deepest level, we are motivated by what we believe to be true and what we ascribe to our identity. Because of this, as the Heath brothers point out, our decision-making is often guided by three questions: "Who am I? What kind of situation is this? What would someone like me do in this situation?"[50] In other words, when we interact with the world, we're gauging a situation through our perceived self-image or set of beliefs and values. How does the information at hand interact with our worldview? This is where we get confirmation bias: if the information we're receiving validates what we currently believe, we accept it. If it questions our worldview, we reject it. And we look for those data points that support our perspective and thus make us feel more confident in our personal identity and our beliefs about how the world works.

When it comes to climate change, someone might answer these questions like this: "I am a conservative. This is a situation in which a liberal is trying to convince me that climate change rhetoric is true. What conservatives are supposed to do in this situation is deny it and say that the science isn't clear or that protecting the environment will cause undue burden to the economy."

I get it. I understand where they're coming from. When I see people with different political views than my own posting their strongly worded opinions on social media, I roll my eyes sometimes. But that moment of dismissal is exactly where the disconnect starts. As essayist Henry Wismayer points out, as humans we have a knee-jerk reaction to someone telling us what to do or how to think.[51] And if there's any hint of self-righteousness in the message, it's even more of an affront. As climate change advocates, how do we expect people to react when we tell them that every aspect of their life is contributing to the destruction of the planet? It's no wonder they're not listening. They're probably thinking, "Here come the liberals with their climate change boogeyman, trying to take our cars and our hamburgers away." Based on how the information is being presented to them, I can see why they feel that way.

In a world in which information is being thrown at us a million miles a minute, most of us don't have time to talk with our friends to find out what's behind their beliefs—let alone people who aren't friends but mere fellow Americans who live in a different part of the country and have a different lifestyle and a different worldview. We've gotten too busy to search for common ground and shared values that could open up productive dialogue. Instead we ignore it. Or we post an angry rant, only further reinforcing the perception of arrogance or self-righteousness. This is partly why we've become more siloed than ever. And as long as this dynamic remains, it will be tough to motivate our collective Elephant, because we won't know how to speak to each other's deep-seated beliefs, values, and emotions and find common ground.

So that's where climate solutions need to start: Sitting at the table, looking each other in the eye, listening, learning, and establishing trust. Seeing this person not as a liberal or conservative, Democrat or Republican, but as an American—as a fellow human being. Like me, this person is concerned about the future of our country. Like me, this person has been through hardship. Like me, they have hopes and dreams. When we stop to look, we can easily see our shared humanity. We can see that we have so much more in common than we might often think.

CHAPTER 3

CLIMATE CONFUSION

Whether you think you can, or think you can't—you're right.

—HENRY FORD[1]

FOR AS LONG AS I CAN REMEMBER, there's been a general consensus that climate change is an issue championed by the Democratic Party. Contrary to popular belief, however, there has been a long history of Republican leaders in the highest positions of leadership who have understood the science of climate change and have championed solutions.

In 1988, NASA scientist James Hansen famously testified before Congress about the dangers of global warming, dramatically increasing broad public awareness of the issue. As the *New York Times* reported,

> Until now, scientists have been cautious about attributing rising global temperatures of recent years to the predicted global warming caused by pollutants in the atmosphere, known as the "greenhouse effect." But today Dr. James E. Hansen of the National Aeronautics and Space Administration told a Congressional committee that it was 99 percent certain that the warming trend was not a natural variation but was caused by a buildup of carbon dioxide and other artificial gases in the atmosphere.[2]

George H. W. Bush, who ran for president that year and was on the campaign trail at the time of Hansen's testimony, responded,

"Those who think we are powerless to do anything about the greenhouse effect forget about the 'White House effect.' . . . As president, I intend to do something about it."[3] He also specified those plans: "In my first year in office, I will convene a global conference on the environment at the White House. . . . The agenda will be clear. . . . We will talk about global warming. We will talk about saving our oceans and preventing the loss of tropical forests. And we will act."[4]

True to his word, President Bush did act on climate change. In 1992, he went to Rio de Janeiro in Brazil for the Earth Summit, where the countries of the world first collectively acknowledged the threat of climate change and signed into effect the United Nations Framework Convention on Climate Change. Upon returning from the conference Bush reported, "We've signed a climate convention. We've asked others to join us in presenting action plans for the implementation of the climate convention. . . . Let me be clear on one fundamental point. The United States fully intends to be the world's pre-eminent leader in protecting the global environment. And we have been that for many years."[5]

In today's political climate, it can seem surprising that a Republican would have been so bold in his goal of tackling climate change and preserving the environment. However, conservatives at that time had a long history of leadership on environmental conservation. Notably, Republican president Teddy Roosevelt was an outdoorsman and an avid lover of nature who established the US Forest Service and the first Federal Bird Reserve, in the process protecting many areas that are now national wildlife refuges and national parks.[6]

In 1962 Rachel Carson, a world-renowned marine biologist for the Fish and Wildlife Service, published *Silent Spring*, in which she documented the devastating effects of pesticides and industrial toxins on the natural environment and birthed the modern environmental movement. On April 22, 1970, Wisconsin senator Gaylord Nelson and student organizer Dennis Hayes set up a national teach-in to raise awareness about what was happening to the environment, which became the first Earth Day. Twenty million people around the country participated in teach-ins, marches, and demonstrations to demand action.[7] Republican president Richard Nixon responded to the public outcry. In 1970 he established the National Environmental Policy Act, the Environmental Protection Agency, and the Clean Air Act and, in 1972, the Clean Water Act.[8] "I think that 1970 will be known as the year of the beginning, in which we really began to move on the problems of clean air and clean water and open spaces for the future generations of America," he proclaimed.[9] So, when President George H. W. Bush insisted that America wanted to continue its legacy of leading the world in environmental protection, his message was in line with a long tradition of Republican support.

The next Republican president was his son, George W. Bush. While the Bush family business was Texas oil, both men clearly understood the science of climate change and made bold promises to take action. In a speech to European leaders in June of 2001, President Bush laid out his thoughts on the matter:

> I've just met with senior members of my administration who are working to develop an effective and science-based approach to addressing the important issues of global climate change.
>
> This is an issue that I know is very important to the nations of Europe. . . . The earth's well-being is also an issue important to America. And it's an issue that should be important to every nation in every part of our world. . . .
>
> There is a natural greenhouse effect that contributes to warming. Greenhouse gases trap heat, and thus warm the earth because they prevent a significant proportion of infrared radiation from escaping into space. Concentration of greenhouse gases, especially CO2, have increased substantially since the beginning of the industrial revolution. And the National Academy of Sciences indicate that the increase is due in large part to human activity. . . .
>
> Our country, the United States, is the world's largest emitter of manmade greenhouse gases. We account for almost 20 percent

> of the world's manmade greenhouse emissions. We also account for about one-quarter of the world's economic output. We recognize the responsibility to reduce our emissions. . . .
>
> . . . My administration is committed to a leadership role on the issue of climate change.[10]

After Hansen's congressional testimony in 1988, the two subsequent Republican presidents agreed that climate change was real and caused by humans, and they said they intended America to lead on the issue. The next Republican to vie for the presidency was Senator John McCain, who ran against Barack Obama in 2008. McCain, a decorated war hero and chair of the Senate Committee on Commerce, Science, and Transportation, did more to call attention to climate change and build bipartisan support for climate solutions than almost anyone at the time, regardless of political party. He repeatedly worked across the aisle with Independents and Democrats to sponsor climate legislation and build coalitions among Republicans to help them pass.

In 2007 CBS's Katie Couric asked McCain if the "risks of climate change are at all overblown":

> MCCAIN: I've been involved in this effort for many years. And we've got to act. And unfortunately, we have not acted either as a federal government or a Congress.
>
> COURIC: Why has it taken so long, Senator?
>
> MCCAIN: Special interests. It's the special interests. It's the utility companies and the petroleum companies and other special interests. They're the ones that have blocked progress in the Congress of the United States and the administration. That's a little straight talk.[11]

After losing the presidential election to Obama, McCain was less vocal about his support for climate change efforts, and many criticized him for being swayed by partisan politics. However, most would agree that among his most notable political achievements over his years of public service were his effort to make climate change a part of the Republican Party's agenda and his tireless work to pass climate legislation.

The next Republican presidential candidate, Mitt Romney, ran against President Obama in 2012. While in the previous election both candidates had run on a platform to solve the climate crisis, the issue was barely mentioned in 2012 (even after Superstorm Sandy had just devastated New York and New Jersey). Oddly enough, the presidential debates didn't raise the climate question, and neither candidate addressed it in detail. As governor of Massachusetts, Romney had a long history of climate action. He ordered the state's first Climate Protection Plan, set vehicle emissions standards, and joined the Regional Greenhouse Gas Initiative (RGGI), which was the first multistate cap-and-trade program to limit CO_2 emissions in the United States.

Earlier, in 2004, Romney had written, "I am proud to announce the Massachusetts Climate Protection Plan, the first in the history of the Commonwealth and among the strongest in our nation. . . . Rather than focusing our energy on the debate over the causes of global warming and the impact of human activity on climate, we have chosen to put our emphasis on actions, not discourse."[12] During the presidential election, his stance on climate change shifted to align with the Republican Party's new emphasis on dismissing climate science. He even proposed rolling back President Obama's climate measures. However, as a senator in 2019, he returned to his original stance. "There's no question that we're experiencing climate change and that humans are a significant contributor to that," Romney told *The Hill*.[13]

The next Republican nominee and president was Donald Trump, who is well known for denying climate change. Like other Republican politicians, he has not expressed a consistent view on the issue. On December 6, 2009, a group of prominent business leaders wrote an open letter to President Obama and the US Congress, which was published in the *New York Times*, urging strong action on climate change at the Copenhagen UN climate negotiations. The letter declared, "We support your effort to ensure meaningful and effective measures to control climate change, an immediate challenge facing the United States and the world today. . . . If we fail to act now, it is scientifically irrefutable that there will be catastrophic and irreversible consequences for humanity and our planet."[14] Included in the list of signatories were Donald Trump and three of his children, Don Jr., Eric, and Ivanka, on behalf of the Trump Organization.

By the time he became the Republican presidential nominee, however, Trump had reverted to the most recent party line: "Climate change is a hoax." Even though countless voters considered the 2016 election critical to climate change, it was, again, barely mentioned during the debates. As president, Trump has mainly been a loud climate denier, aggressively rolling back environmental regulations, pulling out of the Paris Agreement, and putting fossil fuel lobbyists in charge of the EPA and other regulatory agencies. However, he has on occasion admitted that he does believe in the science of climate change. In a *New York Times* interview published November 23, 2016 (weeks after he won the election), President Trump gave a little more insight into his views on climate change:

> JAMES BENNET, editorial page editor: When you say an open mind, you mean you're just not sure whether human activity causes climate change? Do you think human activity is or isn't connected?
>
> TRUMP: I think right now . . . well, I think there is some connectivity. There is some, something. It depends on how much. It also depends on how much it's going to cost our companies. You have to understand, our companies are noncompetitive right now.[15]

Therein lies the predicament.

In these few words, President Trump has summarized the human conundrum concerning climate change. He admits that there is some connection between human activity and climate change. However, the degree to which he believes it's a problem is based not on science but on whether it will negatively impact business. In other words, he's suggesting the physics of climate change "depends on how much it's going to cost our companies."

Another famous climate denier, Republican senator James Inhofe from Oklahoma, made a similar nonsensical argument. He has twice chaired the Committee on Environment and Public Works and once threw a snowball on the Senate floor as part of his argument that climate change is a hoax. In an interview in 2012, he admitted that he originally believed scientists' warnings about climate change but quickly changed his tune when he saw the price tag: "Do you realize I was actually on your side of this issue when I was chairing that committee and first heard about this? I thought it must be true until I found out what it would cost."[16]

This argument is equivalent to not believing the doctor who says you have a broken arm if it means you have to wear a cast all summer.

Both President Trump's and Senator Inhofe's remarks indicate that they don't recognize the logical fallacies contained in their argument. Unfortunately, they're not the only ones whose belief in climate change does a U-turn when it becomes clear that solving climate change will be difficult. This is an interesting insight into our psychology as it relates to climate change: People will choose not to believe in climate change if they think there is no way to solve it. Denial is not a surprising coping mechanism. One of the necessary conditions for people to take action is the belief that they have the skills to help. While climate change denial makes no logical sense, it makes plenty of emotional sense.

EMPOWERMENT

Climate change is the biggest threat humanity has ever faced—and one that threatens all life on our planet. What can an individual in such dire circumstances do? We're often told to do two things: to take individual action (like changing your lightbulbs) and to take political action (like marching in the street or signing petitions). While these are indeed important actions to take, they may seem futile compared to the scope of the problem. Or these actions may not be in line with someone's cultural, political, or social identity. In other words, they may care about climate change, but they don't see themselves as the face-painting, poster-carrying environmental activist type.

Anthony Leiserowitz is the director of the Yale Project on Climate Change Communication, famous for its periodic reports, *Climate Change in the American Mind*, which it publishes with the George Mason University Center for Climate Change Communication. A few years ago, I emailed Leiserowitz with this nagging question: "An area I wonder about is empowerment. For example, when people look at climate change, are they so overwhelmed and feel so helpless to change the situation that somehow their brain decides to deny it or ignore it so that they won't have to feel worried or guilty?" Tony replied, "There's a social science term very related to your concept of empowerment, but it's called 'efficacy'—and has long been understood as a critical factor. People need to know that they are capable of action and that the actions they take will actually make a difference."[17]

His response got me interested in understanding the relationship between efficacy and belief in climate change science, so I started digging into the research. A study done by Yale and George Mason University researchers on attitudes concerning climate changed revealed that "Efficacy—the belief that individuals can make a difference in climate change—positively predicted both belief and attitudes. . . . It is thus highly likely—though perhaps at first counterintuitive—that enhancing a sense of personal empowerment may be an effective communication strategy to spur belief in climate change."[18] In other words, an individual's sense of personal empowerment (the belief that they can make a difference in climate change) *predicted* whether they believed in climate change science at all. In other words, the feeling of empowerment preceded their acceptance of the science. While this relationship between empowerment and belief sounds paradoxical, it helps explain the varying degrees of acceptance of climate change across the country.

Additionally, the study highlights factors that stand in the way of belief: "An efficacy message also offers a potentially useful counterpoint to traditional climate change messages that have focused on government solutions, which tend to play less well among Republican audiences."[19] If one of your main values as a Republican is to have a small government, and you get the message that the only way to stop climate change is through big government, you're likely to simply reject the science altogether. When climate change activists promote government solutions as the only way to solve climate change, they play into the skeptic's suspicion that climate change is a conspiracy intended to increase government control in our lives.

In order to get energized about solving climate change, then, the public needs to feel empowered to make a difference. In between individual action and advocating for legislation at the federal level, there are efforts we can take at the *community level.* At the community level our collective actions can produce *tangible results,* as well as practical benefits, giving us the greatest sense of efficacy and empowerment. These local actions have the best chance to change the conversation around climate change in America.

And yet, we don't necessarily need all climate skeptics' beliefs to change. If efforts to combat climate change locally—for example, putting up solar panels on the local school—are creating tangible benefits for the community, such as saving money and creating jobs, then people will get involved and support the effort regardless of their belief in climate change. And that has the potential to spread community to community, county to county, and state to state.

UNLIKELY ALLIES

Along these lines, there is solid evidence that clean energy sources are good for national security, a crucial issue for conservatives. A 2003 Pentagon study determined that climate change significantly affects our ability to defend ourselves and thus "should be elevated beyond a scientific debate to a US national security concern."[20] Approaching two decades later, our leaders are still debating. In November 2018, Trump's White House released volume 2 of the Fourth National Climate Assessment, which the federal government is required by law to publish every four years. According to the *New York Times*, "In direct language, the 1,656-page assessment lays out the devastating effects of a changing climate on the economy, health and environment, including record wildfires in California, crop failures in the Midwest and crumbling infrastructure in the South. Going forward, American exports and supply chains could be disrupted, agricultural yields could fall to 1980s levels by midcentury and fire season could spread to the Southeast, the report finds."[21]

Trump denounced the findings of the report that his own administration produced, using the research of federal scientists across multiple federal agencies:

> REPORTER: Mr. President, have you read the climate report yet?
> TRUMP: I've seen it. I've read some of it and it's fine.
> REPORTER: They say the economic impact would be devastating—
> TRUMP: Yeah, I don't believe it.
> REPORTER: You don't believe it?
> TRUMP: No, no, I don't believe it.[22]

In contrast, many national and local Republican leaders have publicly acknowledged the reality of human-caused climate change and are taking action to solve it, including mayors, military leaders, business leaders, and leaders of faith communities. Here are a few notable quotes from additional Republican Party leaders who have long acknowledged the reality of climate change and made calls for action:

Senate Majority Leader Mitch McConnell (R-KY):

I do [believe in human-caused climate change]. The question is how do you address it. . . . The way to do this consistent with American values and American capitalism is through technology and innovation.[23]

Senator Lindsey Graham (R-SC):

I believe climate change is real. I also believe that we as Americans have the ability to come up with climate change solutions that can benefit our economy and our way of life. The United States has long been a leader in innovation. Addressing climate change is an opportunity to put our knowledge and can-do spirit to work to protect the environment for our benefit today and for future generations.[24]

I can't imagine a nominee for either major party arguing to the public that climate change is not real, and man is not contributing to it. . . . If they take that position, the public is going to really question their judgment.[25]

Former Speaker of the House John Boehner (R-OH):

I think most members think that climate change is a serious issue that needs to be addressed.[26]

The fact is, is that we have had climate change. Clearly, humans have something to do with it, and we ought to begin reducing our CO_2 emissions.[27]

Former Speaker of the House Newt Gingrich (R-GA):

Our country must take action to address climate change.[28]

As a conservative, I think you ought to be prudent, and it seems to me that the conservative approach should be to minimize the risk of a really catastrophic change.[29]

I'm trying to say to the right the environment's too important to neglect. The issues are too serious to walk away from, and therefore you've got to drop just screaming "No," and you got to show us what the right solutions are from your standpoint.[30]

Former mayor Rudy Giuliani (R-NY):

I do believe there's global warming, yes. . . . The overwhelming number of scientists now believe that there is significant human cause. . . . The issue of global warming is no longer a Democratic issue only . . . but a Republican issue as well.[31]

If all of these GOP leaders, including every Republican president or nominee since George H. W. Bush, have acknowledged the science of climate change, why is climate change still considered a partisan issue?[32] What is holding us back from solutions-based dialogue?

This is going to be a big shocker. Could it be . . . ?

Yes, you guessed it. Money!

MONEY IN POLITICS

One day, a friend of mine who worked in Washington, DC, explained her understanding of money in politics, and I'll never forget it. She explained how a lobbyist will walk into a politician's office and clearly lay out the issues that they care about. Then they say, "Look, I've got a whole bunch of money that I plan to use this campaign season. I can use it to support you and your campaign, or I can use it to support your opponent." This description of corruption in politics may seem a bit simplistic, but frankly, I doubt it's much more complicated than that. Money in politics has been a problem in this country, and in most countries, since the invention of money. And when the fossil fuel industry, the most profitable industry in the history of money, as Bill McKibben likes to say, is facing an existential threat to their underlying business model, it's not surprising that there's so much money being thrown at the issue.

According to the Center for Responsive Politics, the oil and gas industry alone contributed over $80 million during the 2018 elections, with 87 percent going to Republican candidates and the Republican Party. In 2016, it was over $100 million.[33] Electric utilities spent over $50 million on these same two elections, over 62 percent of which went to Republicans.[34] Coal mining industries spent over $20 million during the same time frame, over 94 percent of which went to Republicans.[35]

Interestingly though, it's not as one-sided as it may seem. In the 2016 elections, the second and third top recipients (after Republican senator Ted Cruz) of oil and gas money were Donald Trump and Hillary Clinton. They received $1,109,893 and $986,622 respectively.[36] The fossil fuel industry isn't taking any chances. It is buying off both sides to ensure that the scourge of climate policy won't affect their business.

While it may seem shocking that the fossil fuel industry would be so brazen in its attempts to influence politics and public perception regarding the harmful effects of its products, it shouldn't be. We've seen this happen countless times before, the most recent example being the harmful effects of smoking. In fact, the fossil fuel industry took a page right out of the tobacco industry's playbook on how to sow doubt in the public's mind. They even hired the same consultants to help them create misinformation campaigns.[37]

SOWING DOUBT

In 2015 *InsideClimate News* and the *Los Angeles Times* broke a story that ExxonMobil and other leading fossil fuel companies knew about the role their products played in causing climate change as far back as the 1960s. In 1965, President Lyndon B. Johnson's Science Advisory Committee authored a comprehensive report called *Restoring the Quality of Our Environment*, which acknowledged the threat caused by increased carbon dioxide in the atmosphere and admitted that it could lead to melting ice caps, rising sea levels, acidification of water sources, and "produce measurable and perhaps marked changes in climate, and will almost certainly cause significant changes in the temperature and other properties of the stratosphere."[38] The same year, Frank Ikard, president of the American Petroleum Institute (API), discussed the findings of the report at the annual API conference:

> This report unquestionably will fan emotions, raise fears, and bring demands for action. The substance of the report is that there is still time to save the world's peoples from the catastrophic consequence of pollution, but time is running out.
>
> One of the most important predictions of the report is that carbon dioxide is being added to the earth's atmosphere by the burning of coal, oil, and natural gas at such a rate that by the year 2000 the heat balance will be so modified as possibly to cause marked changes in climate beyond local or even national efforts. The report further states, and I quote: ". . . the pollution from internal combustion engines is so serious, and is growing so fast, that an alternative nonpolluting means of powering automobiles, buses, and trucks is likely to become a national necessity."[39]

Over half a century has passed since our government and fossil fuel industry leaders realized exactly what was happening and how serious this existential threat was. We have little to show for it.

The American Petroleum Institute started doing its own research very early. A study by the Stanford Research Institute prepared for the American Petroleum Institute in 1968, like the earlier Science Advisory Committee report, laid out clearly that rising CO_2 could result in increases in temperature at the earth's surface, which could lead to melting ice caps, rising seas, and potentially serious environmental damage worldwide.[40] It even predicted that by the year 2000 we would have a CO_2 concentration in the atmosphere of 400ppm (which was only off by a few years). Exxon, in particular, spent millions studying the issue internally and was well aware of the impacts CO_2 was having on the climate. According to *InsideClimate News* reporter Neela Banerjee, "By the mid-1970s, Exxon began to take carbon pollution seriously. In July 1977, James Black, a senior scientist at Exxon, told top executives that carbon dioxide emissions from burning fossil fuels would warm the atmosphere and endanger human life."[41]

Nonetheless, Exxon chose to tell a different story to the public. The *Los Angeles Times* reported that, a month after NASA scientist James Hansen warned Congress about climate change in 1988, Exxon issued an internal memo advising that they "emphasize the uncertainty" in the scientific data about climate change.[42] Rather than doing something about the problem or telling the public what it knows, Exxon has spent the last five decades running a disinformation campaign that has delayed political action. They also found in their research the possibility of increased profits: they measured how much longer the arctic sea would be open to drilling because there would be less sea ice each year, and they started building taller oil rigs to prepare for sea level rise.[43]

This strategy of misinformation extended to the government. Political consultant Frank Luntz, who famously wrote the playbook for Republican messaging on climate change, was an advisor to President George W. Bush.[44] Despite Bush's lofty climate language soon after he was elected, Luntz advised the administration to follow the same strategy as the fossil fuel industry. In an internal memo to the Bush administration, Luntz wrote, "Voters believe that there is no consensus about global warming within the scientific community.

Should the public come to believe that the scientific issues are settled, their views about global warming will change accordingly. Therefore, you need to continue to make the lack of scientific certainty a primary issue in the debate."[45]

In order to sow doubt in the minds of Americans, the fossil fuel industry knew exactly where to turn to: the same consultants that helped cast doubt over the harmful impacts of tobacco smoking (detailed in Naomi Oreskes and Erik Conway's book *Merchants of Doubt*). *Scientific American* reports that as early as the 1950s, the tobacco and oil industries "shared scientists and publicists to downplay dangers" of their products to the public.[46] "From the 1950s onward, the oil and tobacco firms were using not only the same PR firms and same research institutes, but many of the same researchers," remarked Carroll Muffett, president of the Center for International Environmental Law, a Washington, DC–based advocacy group.[47] They also funded think tanks for decades whose sole purpose was to share misinformation about climate change with the public. Banerjee's 2017 *InsideClimate News* story further reported,

> Hundreds of millions of dollars from corporations such as ExxonMobil and wealthy individuals such as the billionaires Charles and David Koch have supported the development of a sprawling network, which includes Heartland and other think tanks, advocacy groups and political operatives. They have cast doubt on consensus science, confused public opinion and forestalled passage of laws and regulations that would address the global environmental crisis. It is one of the largest, longest and most consequential misinformation efforts mounted against mainstream science by an industry.[48]

The fossil fuel industry of course does more than just spread misinformation to the public and give large financial contributions to candidates running for office. It also uses its money to influence local and state policies around the country through a lobbying group, the American Legislative Exchange Council (ALEC). Even though they now acknowledge the science behind climate change,

Shell, Chevron, and ExxonMobil continue to fund ALEC, which uses the money to introduce cookie-cutter legislation that weakens climate and energy policies and fights state renewable energy standards.[49] Unfortunately, they're very good at it, and the lobbyists who testify on behalf of ALEC are professionals. They're hired guns who go state to state, fighting against clean energy and climate policy, paid by the fossil fuel industry.

Since 2015 the top five oil companies (BP, Chevron, ExxonMobil, Shell, and Total), which all publicly supported the Paris Agreement to address climate change worldwide, now collectively spend over $200 million per year on lobbying to prevent climate solutions. They also spend an additional $195 million "on branding campaigns that suggest they support an ambitious climate agenda."[50] Yet all of this is pennies compared to their steady focus on business as usual. To put things into perspective, the same five oil companies in 2018 collectively made $230 million *per day*.[51]

As I write this, I think about all the executives who have known that their product would cause climate havoc and proceeded to pursue their line of business regardless, lying to the public and buying politicians to protect their profit margins. I can't help but wonder, when they play with their children and grandchildren, don't they ever ask themselves, What kind of world am I leaving them? I can only presume that when you're one of many staff at a big corporation, perhaps peer pressure influences your behavior. Perhaps they feel as if they have no agency: *If no one else is saying anything, I probably shouldn't either*. Or, maybe, they're just so fixated on making money that they've gotten used to ignoring it. I really don't know.

CLIMATE IN THE NEWS

Besides the industry-funded conservative think tanks like the Heartland Institute, which launched a billboard campaign in 2012 to compare believers in global warming to "'murderers and madmen' such as the Unabomber, Charles Manson and Osama bin Laden,"[52] where else has the Republican mainstream audience been hearing climate misinformation?

The well-oiled media machine, of course. (Pun intended.)

A George Mason University study on the effect of media on beliefs about climate change found the following:

> Research into the relationship between cable television news viewing patterns, political partisanship and belief in climate change finds evidence for a strong media effect, with high levels of Fox News viewership among Republicans predicting lower levels of belief in climate change. Feldman et al. also note substantial crossover viewing by Republicans watching politically moderate to liberal outlets CNN and MSNBC. Republicans who frequently viewed these stations were more likely to agree that climate change is in fact occurring.[53]

In addition, a March 2019 study revealed that 62 percent of Americans who don't watch Fox News agree that climate change is caused largely by human activity, compared to only 12 percent of Fox News viewers.[54]

How does this disparity happen? Because media coverage of climate change is so skewed. For example, NASA points out that multiple studies demonstrate that 97 percent of climate scientists or more agree that climate change is real and is caused by humans.[55] However, when the United Nations Intergovernmental Panel on Climate Change came out with its 2013 report, Fox News coverage of the report gave 69 percent of airtime to climate doubters and only 31 percent of airtime to those in agreement with the vast majority of climate scientists and the findings of the report. The report showed that even media sources considered more moderate, such as *Bloomberg* and the *Los Angeles Times*, were guilty of giving climate deniers more time and space than they were due.[56]

Of course, Fox isn't the only culprit. Media Matters points out that in March 2019, when massive floods hit the Midwest, "ABC, CBS, and NBC completely failed to mention climate change."[57] As a senior climate writer at Media Matters describes it:

> A bomb cyclone of "historic proportions" began raging across the Midwest on March 13. It unleashed a torrent of wind, snow,

> and rain that caused unprecedented flooding in Nebraska as well as floods in Iowa, Minnesota, Missouri, South Dakota, and Wisconsin, resulting in at least four deaths and $3 billion in losses. The floods destroyed hundreds of homes and affected millions of acres of farmland.[58]

Meanwhile, other stations appropriately pointed out that the National Climate Assessment and other scientific sources suggest that these are the exact types of dramatic events caused by climate change.

I'm not sure why the media coverage of climate change is so skewed, but I know it's not helping. Could it be that advertisers—for example, car companies—don't want media outlets talking about it because it might hurt their businesses? Or that their parent corporations are allied with fossil fuel companies? Or is it just their brand to align with the conservative politicians who have been bought by the industry? Perhaps they find that giving people a false hope that climate change isn't real is good for ratings. Or perhaps taking a contrarian approach to what other news outlets are saying, no matter how off-base their claims are, and marketing themselves as truth tellers, is good for ratings. Whatever the case may be, the media (particularly the conservative-leaning media sources) are proving to be an incredible impediment in the effort to unite our country around climate solutions.

However, despite all the forces stacked against them, there are still a number of conservatives who see through the politics, who know the stakes are high, and who are putting their effort into changing the conversation. These unlikely climate heroes, who we'll hear from in the next chapter, demonstrate what it means to have climate courage.

CHAPTER 4

CONSERVATIVES FORGING A PATH

Public sentiment is everything. With public sentiment, nothing can fail; without it, nothing can succeed.

—ABRAHAM LINCOLN[1]

I'VE GOT SOME GOOD NEWS, and I've got some great news. I'll start with the good news.

As we learned in chapter 3—despite the fact that for over thirty years every Republican president or presidential nominee has acknowledged the science behind climate change—due to the massive misinformation campaigns paid for by the fossil fuel industry and promulgated by think tanks, lobbyists, Republican politicians indebted to the industry, and conservative media, there still is a partisan divide across the country when it comes to acknowledging the science of climate change. However, this divide is actually getting a lot narrower.

Here are some of the highlights from a 2019 survey, *Climate Change in the American Mind*, conducted jointly by the Yale Program on Climate Change Communication and George Mason University Center for Climate Change Communication (I have italicized text for emphasis):

- Seven in ten Americans (72%) think global warming is happening, an increase of ten percentage points since March 2015.

- Only about one in eight Americans (12%) think global warming is not happening.
- *Americans who think global warming is happening outnumber those who think it isn't by more than a 6 to 1 ratio.*
- About six in ten Americans (59%) understand that global warming is mostly human-caused.
- However, only about one in five (22%) understand how strong the level of consensus among scientists [about the reality of climate change] is (i.e., that more than 90% of climate scientists have concluded that human-caused global warming is happening).
- Two in three Americans (67%) say the issue of global warming is either "extremely," "very," or "somewhat" important to them personally, while one in three (33%) say it is either "not too" or "not at all" personally important.
- More than six in ten Americans (64%) think global warming is affecting weather in the United States.[2]

The following results are from the *Politics and Global Warming, November 2019* survey conducted by the same teams at Yale and George Mason.

- *Most registered voters (73%) think global warming is happening*, including 95% of liberal Democrats, 89% of moderate/conservative Democrats, and *66% of liberal/moderate Republicans. Only 41% of conservative Republicans think global warming is happening.*
- A majority of registered voters (59%) think global warming is caused mostly by human activities.
- *Two in three registered voters (66%) are worried about global warming*, including 94% of liberal Democrats, 88% of moderate/conservative Democrats (an increase of 20 percentage points over the past five years), and 53% of liberal/moderate Republicans. Only about one in four conservative Republicans (26%) are worried.[3]

Long story short, the vast majority of Americans believe climate change is happening (72 percent) and that it's human caused (59 percent). It's sad that only 22 percent of Americans realize that the vast majority of climate scientists (97 percent) agree on the science, but that's increasing. So while there's still some room for education on the issue, we're certainly moving in the right direction.

If you are an elected official reading this, what should make you straighten up in your chair is the fact that two in three registered voters (66 percent) are worried about global warming. *Worried.* That means that they're looking to you to help solve the problem. And as an elected official, you'd better acknowledge the science and work on a plan to solve it. Fast. Gone are the days when you could say, "I'm not a scientist," and punt on the issue. That won't fly in today's America, where the vast majority of Americans understand the problem fairly well.

Now while these trends are all fairly good news, here's the great news: We can *all* agree on one thing—Americans want a swift transition to a 100 percent clean energy economy. Republicans and Democrats alike *all* get it. It's time to move off fossil fuels. In other words, while it's fantastic that more people understand the science of climate change, in terms of taking action, we all already agree on the solutions, for a whole host of reasons, not just climate change.

Some highlights from the December 2018 report *Energy in the American Mind*, also produced by Yale and George Mason (I've italicized text for emphasis):

- A majority of Americans (58%)—including three in four Democrats (75%)—think policies intended to transition from fossil fuels to clean, renewable energy will improve economic growth and create new jobs. Only 18% of Americans . . . think such policies will *reduce* growth and jobs. More Republicans think such a transition will improve economic growth (39%) than reduce it (31%).

- Americans' most important reasons to support a transition to 100% clean, renewable energy are reducing water pollution (75%), reducing air pollution (74%), and providing a better life for our children and grandchildren (72%). [Note, climate change is not one of them—and that's OK!]
- About seven in ten Americans (71%; including 87% of Democrats and 51% of Republicans) think clean energy should be a "high" or "very high" priority for the president and Congress. Very few Americans (7%, including 3% of Democrats and 11% of Republicans) think it should be a "low" priority.
- Between 2013 and 2018, support for funding renewable energy research increased by 15 percentage points among registered voters (from 73% to 88%). *This growing support was primarily driven by a large shift among conservative Republicans, whose support increased by 30 points, from 50% to 80%.*
- Between 2013 and 2018, support for tax rebates for energy-efficient vehicles and solar panels increased by 12 percentage points among registered voters (from 73% to 85%).[4]

Here, in my opinion is the most exciting few lines in the report, and perhaps even in this book.

- *A large majority of registered voters (85%)—including 95% of Democrats and 71% of Republicans—support requiring utilities in their state to produce 100% of their electricity from clean, renewable sources by 2050. Nearly two in three conservative Republicans (64%) support this policy.*[5]

Mic drop.

Not sure if you all caught that. Eighty-five percent of registered voters want to see a transition to 100 percent clean, renewable energy. So . . . note to politicians of every stripe: clean energy is what the people want across the political spectrum. Everything else is old news. They want it for multiple reasons, and these reasons include its value to our economy. Gone are the days of "all of the above" energy policies. Gone are the days of voters believing false claims that switching to renewable energy will harm the economy. The American voter isn't falling for it anymore. Better get with the program.

Let's reflect for a moment on the rarity of what this report reveals. What else do 85 percent of Americans agree on? A bipartisan clean energy push is perhaps one of the few things that could really unite us all behind a common cause.

CLIMATE COURAGEOUS CONSERVATIVES

Since the environment and energy have been such divisive issues in recent years, it's not surprising that Republican politicians aren't yet cheerleaders for climate solutions and clean energy, even though the vast majority of voters are. It appears that there's a bit of lag time before the leaders catch up with the rest of us.

Thankfully, there are a number of conservative politicians and thought leaders who have been brave enough to take a stance on climate change in spite of the political atmosphere. We really owe these unlikely climate heroes a debt of gratitude for showing what's possible, even when it meant facing public opposition, and in some cases, losing their seats in Congress. These brave souls exemplify climate courage.

The Governator

I'm going to start with a Republican climate hero who needs no introduction. Arguably the greatest bodybuilder of all time. The Terminator. The Governator. The Last Action Hero. And the man who, while governor of California, signed into effect the first statewide greenhouse gas cap-and-trade program in the US designed to curb global warming. Of course, I'm talking about the one and only Arnold Schwarzenegger.

As a Republican, Schwarzenegger took a traditional stance on a lot of issues. He's a believer in free enterprise, reduced government regulation, low taxes, and a strong military.[6] But what stood out during his tenure were his groundbreaking efforts to ameliorate climate change, spur clean energy development, and, in the process, grow the economy of his state. What makes Schwarzenegger's story so exciting is that not only was he the only *Republican* governor who was pushing climate and clean energy legislation that bold; he was the only governor *period* to do so.

Granted, California has a history of being at the cutting edge of environmentally conscious energy policy, such as allowing for utilities to promote energy efficiency by "decoupling" the amount of electricity sold from the profits they make. But it still took a lot of political guts as a Republican to push for the first greenhouse gas cap-and-trade policy in the nation in 2006. Schwarzenegger stuck his neck out on a policy that barely passed. Of all Republicans voting in the legislature on Assembly Bill (AB) 32, he got only one Republican vote. Only one.

Thankfully, his policies now speak for themselves. California produces more clean energy than almost any state in the country and is one of the leading epicenters around the world of clean energy innovation and job creation. Overall, the state's economy has continued on an upward trajectory since implementation of the law. Governor Schwarzenegger, in a 2019 op-ed, pointed out that California's environmental standards boost the economy rather than hinder it:

> Last year, the U.S. economy grew by 2.9 percent. California's economy, with our supposedly crippling regulations, grew by 3.5 percent. We've outpaced the nation's economic growth even as we've protected our people.
>
> Our success is built on our consistency. Ever since Reagan, each governor has continued the legacy of moving toward a clean energy future.
>
> . . . You can't just erase decades of history and progress by drawing a line through it with a Sharpie.[7]

He wrote the editorial in direct response to the Trump administration's attempt to roll back California's automobile standards.

Schwarzenegger has been keen to point out that environmental conservation is squarely in line with conservative values: to take the more prudent, less risky, and safer course of action. In 2007 he spelled it out nicely in a relatable metaphor: "It is like when my child is sick and has a huge fever and I go to 100 doctors and 98 doctors say this child needs immediate medical care and two say, 'No, it's OK, just go home and relax.' . . . I go with the 98."[8]

Well said, Governor.

Mayor Dale Ross, Georgetown, Texas

When I first moved to the Bay Area, I met a solar installer who said that solar panels were for "sandals and candles" people. I had a chuckle. The first city in America to meet 100 percent of its power needs with renewable energy was Burlington, Vermont.[9] No surprise there. Plenty of sandals and candles. But the second city in the US to run on 100 percent renewable energy? Georgetown, Texas, led by Republican mayor Dale Ross—with a population of twenty thousand more people than that of Burlington, by the way.[10]

Mayor Ross is a certified public accountant and has his own accounting business. He's also a pretty conservative guy. According to Dan Solomon, who profiled Ross in *Smithsonian*, "His priorities are party staples: go light on regulation, be tough on crime, keep taxes low."[11] In fact, among his accolades as mayor, he's able to boast that the city has the lowest effective tax rate in Central Texas.[12] As Ross himself puts it, "Georgetown . . . is the reddest city in the reddest county in Texas, and I'm a conservative Republican."[13] What would compel them to go 100 percent green? "No, environmental zealots have not taken over our city council, and we're not trying to make a statement about fracking or climate change. Our move to wind and solar is chiefly a business decision based on cost and price stability," Mayor Ross wrote in an op-ed.[14]

Well, as it turns out, this conservative puts his fiscal responsibility above partisanship. While going 100 percent renewable might not be a high-ticket item on the Republican platform, it was the most cost-effective way to power the city. The City of Georgetown owns and operates its own municipal utility, which means it can choose who it gets its power from and can negotiate directly with power providers. The natural gas providers could only offer contracts that would lock in a price for seven years, but the wind energy provider was able to guarantee low power costs for twenty years and the solar energy provider for twenty-five years. It was clear to Mayor Ross that putting the financial well-being of his constituents first meant going with renewables.

Now keep in mind, this isn't some sleepy town in Texas. According to the mayor it's the "fastest-growing large metro area in the U.S.," and therefore "we needed more capacity to meet projected growth."[15] In addition to recognizing the cost benefit, he also realized that the move would help the city conserve precious water resources: "Drought conditions and half-empty reservoirs have been common in Texas in recent years. Traditional power plants making steam from burning fossil fuels can use large amounts of water each day. Our move to renewable power is a significant reduction in our total water use in Georgetown."[16]

Ross also sees the switch to clean energy helping to attract business to his city. "Many companies, especially those in the high-tech sector, are looking to increase green sources of power for both office and manufacturing facilities," he says. "Our 100 percent renewable energy can help those companies to achieve sustainability goals at a competitive price without the burden of managing power supply contracts."[17] He's right. As tech companies look to move their offices, low-cost clean energy is certainly a strong competitive benefit for a city like Georgetown.

Mayor Ross is not just a smart mayor. He's also courageous. Like Governor Schwarzenegger, here is a leader who did what was right for his city despite the optics and political backlash he might face for what could be considered a liberal move. "You think of climate change and renewable energy, from a political standpoint, on the left-hand side of the spectrum, and what I've done is toss all those partisan political thoughts aside," Mayor Ross says. "We're doing this because it's good for our citizens. Cheaper electricity is better. Clean energy is better than fossil fuels."[18]

Debbie Dooley and the Green Tea Party

Another unlikely clean energy and climate solutions champion is Debbie Dooley, one of the founders of the Tea Party, hailing from Atlanta, Georgia. Dooley has a surprising story of how she came to be a clean energy spokesperson. In the mid-2000s Dooley started to pay attention to what was happening with her utility, Georgia Power. They were planning to build two new nuclear facilities, with costs expected to exceed the original estimates, and they were going to make a guaranteed profit on the whole thing. Who was paying for it? The rate-payers of course, and in advance. As a founding member of the Tea Party who supports free markets, she realized that utilities are government mandated monopolies—the opposite of a free market. To insert some competition into the equation, she became an advocate for solar.

In order for solar to play a role, though, laws needed to change. Realizing that she couldn't do that on her own, Dooley reached out to an unlikely ally: the country's largest and oldest environmental group, the Sierra Club. Together they fought for strong solar policy in Georgia and won. They fittingly called their group the Green Tea Coalition.

They didn't just stop in Dooley's home state of Georgia, though. This band of solar allies then took the fight to Florida Power and Light. Part of what made the effort so successful was that this was a coalition across party lines. Bipartisan support may be hard to get on a lot of issues, but when it comes to solar, conservatives and liberals agree. The fuel source—sunshine—is free and limitless. It's clean, which is good for our air and water, and it creates jobs. And as Dooley puts it, it's about energy freedom and energy independence: "Solar empowers the consumer and the individual. These giant monopolies want to take away that consumer choice unless they can control it. They are looking at their profit margins, not the best interests of ratepayers."[19]

Dooley also knows that because the power grid is so centralized, it's a national security issue. In 2014 a group of snipers shot at a City of Palo Alto Utilities substation for nineteen minutes. It was described as "the most significant incident of domestic terrorism involving the grid that has ever occurred" by the chairman of the Federal Energy Regulatory Commission. These snipers successfully knocked out seventeen transformers, which were down for a month.[20] "Rooftop solar makes it harder for terrorists to render a devastating blow to our power grid," she said. "There's nothing more centralized in our nation. If terrorists were able to take down nine key substations, it would cause a blackout coast to coast."[21]

Dooley says the utilities from Wisconsin to Georgia that are charging fees to solar customers are essentially imposing a tax. "You can call it whatever you want to, it is a tax. I think that is totally ridiculous."[22] Behind these efforts around the country, as we learned in chapter 3, is the lobbying group the American Legislative Exchange Council (ALEC), which, the *Guardian* reported in 2013, is "an alliance of corporations and conservative activists . . . mobilizing to penalize homeowners who install their own solar panels" as part of "a sweeping . . . offensive against renewable energy."[23]

Carolyn Kormann, writing in the *New Yorker*, had this to say:

> Utility companies are not wrong to fear rooftop solar. Its popularity, if unchecked, will certainly cut into their profits and, perhaps, into their budget for maintaining the electrical grid. Hence an ALEC campaign, revealed by the *Guardian* last winter, to promote legislation to penalize individual homeowners who use rooftop solar and to label them "free riders." This effort is part of a broader campaign against renewable energy—solar, wind, biomass—that ALEC, whose leading members include such fossil-fuel giants as ExxonMobil and Peabody Energy, has been conducting for years. In 2012, according to the *Guardian*, "The group sponsored at least 77 energy bills in 34 states," many with the goal of blocking renewable-energy efforts and weakening clean-energy regulations. The model legislation developed by ALEC in this area is called, of all things, the Electricity Freedom Act.[24]

It's not lost on Debbie Dooley that it's rare for someone of her conservative political views to be a strong supporter of solar. But she knows it's all in how the issue is framed: "I believe conservatives who believe in the free market would be receptive to the right message. If you go out and say we need solar because of climate change and you hate coal, that's the wrong message. If you go out and hold elected officials accountable for supporting these monopolies, that's something conservatives will respond to."[25] She has become something of a spokesperson for solar among conservatives. "You would be really surprised about conservatives," she says, "how many are asking me about solar panels. It empowers the individual and it's good for the environment. To me conservation is conservative. . . . You can put solar panels on your rooftop, as long as you have daylight and a battery you can power yourself."[26]

Dooley also raises the question of subsidies—something many liberals concerned about climate change have been pointing out for years. While it is often made to sound as if renewables are only cost competitive due to government subsidies, just the opposite is true—it's only because of the subsidies dirty energy technology has received for generations that it appears to be so cheap. Let's face it, for renewable energy, the cost of the fuel source is free. There's really nothing that can compete. Dooley says,

> I've advocated from the beginning to cut out massive subsidies that all energy forms have received. Some of the same conservatives that point a finger at [solar panel company] Solyndra, they failed to point out the massive subsidies coal and nuclear have been receiving since the 1930s. That smacks of hypocrisy. If you're going to complain about subsidies for wind or solar, why are we still subsidizing nuclear and coal? Subsidies are government's way of picking winners and losers.[27]

For reference, the fossil fuel industry receives about twice as much in direct subsidies as renewable energy does in the US. In 2016, fossil fuels received over $20 billion in federal support, while renewables received under $11 billion.[28] However, a study by the International Monetary Fund found that when you also factor in the full value of indirect subsidies, coal, oil, and gas received $649 billion of support in 2015 in the US. To put that in perspective, that's ten times more than we spent on education, and even more than the

$599 billion spent on US defense, which accounted for 54 percent of the discretionary budget that year. Dooley further explains:

> I talked at the Tea Party convention and said government should stop picking winners and losers. I got thunderous applause. People are really receptive to the right message when you lay the facts out.
>
> I've been attacked by Koch brothers–affiliated groups. [Some] said I was a fake Tea Party person. I just laughed. I am a conservative. There is no way they can paint me as a liberal, even as a moderate—I am a right-wing conservative.
>
> But I'm extremely passionate about alternative energy. I'm a grandmother, I became a grandmother in 2008 with the birth of a grandson who is the light of my life. I want him to have a clean world, I want him to have parks and green space. I don't want him to have to worry about terrorist attacks. I want him to be able to breathe clean air. I want him to be able to generate his own electricity if he so chooses.
>
> If he says, "I want to generate my own power and be self-sufficient," I want him to have the ability to do it.[29]

Dooley knows that conservatives have been "brainwashed for decades into believing we're not damaging the environment."[30] So when she speaks to conservatives about the benefits of renewable energy, she approaches the issue through things she knows they care about: energy choice, energy freedom, national security. If you first connect with people on shared values, then "you have a receptive audience and they will listen to you," she points out. But "if you lead off with climate change, they're not going to pay a bit of attention to anything else you say." (In other words, their Elephant emotional mind will quickly write you off.)

And I think that's pretty good advice for the rest of us too. When we're speaking to an elected official, a neighbor, or a family member, can we first find common ground? What common values do we share? What perspectives and points of view do we have that are similar? Can we use shared language that speaks to those values as we discuss our support for climate solutions?

While it may seem trivial, one of the most important things we can do as Americans who are concerned about climate change is talk about it more. Making it OK to talk about something is the first step toward more shared understanding, and hopefully more collective action.

Bob Inglis—From Climate Skeptic to Advocate

Yet another unlikely climate hero is Bob Inglis, a former Republican congressman from South Carolina. For many years, Inglis served his district as an exemplary conservative member of Congress. In the 1990s, Inglis thought that climate change was "hooey." Describing that viewpoint on a speaking tour in 2018, he told students at the College of Charleston, "If Al [Gore] was for it, I was against it. . . .That's who I was for six years."[31]

It wasn't until Inglis was running for reelection and his eldest son was voting for the first time that a fateful conversation made it clear to him that he needed to examine his thinking. His son told him that only if he cleaned up his act on the environment would he get his vote. That made him question his assumptions. He joined the House Sciences Committee and learned what the scientists were discovering. He visited Antarctica as part of a congressional delegation and saw with his own eyes the mark of carbon dioxide variations in the deep polar ice cores that researchers extracted, which began in the Industrial Age.[32]

On another congressional trip Inglis went to the Great Barrier Reef and learned how the spiking ocean temperatures were bleaching and killing the majestic coral. As he listened to the oceanographer describe the beauty of the reef's natural wonder he believed the scientist "was preaching the Gospel. . . . I could see it in his eyes, I could hear it in his voice. I could see it written all over his face—that he was worshiping God in what he was showing me."[33] And thus he began a process of spiritual reflection. He thought about what he describes as "God's message of love," and he applied it to the future generations he would never have a chance to meet.

Inglis analyzed the available data and understood that climate change was real and being caused by humans. As a conservative who subscribes to the precautionary principle, he knew we needed to do something about it.

In his 2017 TED Talk, Inglis describes how he reoriented his entire perspective on the situation and began looking for conservative approaches to solve the problem. "You bet free enterprise can solve climate change," he says. "Milton Friedman would say to tax pollution rather than profits."[34] Together, these forces helped a conservative member of Congress do a 180 on climate change. Not only did his beliefs change but so did his actions. In 2009 Inglis introduced a carbon tax in the House of Representatives that taxed combustible fossil fuels and carbon intensive imports and would redistribute those tax revenues to Americans by reducing payroll taxes by the same amount. He saw this as a great opportunity to help America become a leader in clean energy and climate solutions in line with conservative values. Inglis wanted his party to be the Grand Opportunity Party, not the Grumpy Old Party.

Unfortunately, his party hadn't gone through the same transformation that he had. His policy didn't gain traction, and during the next election he lost to his opponent, who stuck to the party line. While it cost Inglis his seat in Congress, he knows what he did was right: "Given a choice between the law of love and the law of politics, I know I chose the better."[35]

Thankfully, Bob Inglis didn't stop there. He knew that he was on to something and that he needed to get his message out. Inglis has gone on to start an organization, RepublicEn, that promotes conservative policy solutions to climate change. He believes that conservative free market policies hold the key to stopping climate change. Thankfully, this message is reaching young Republicans across the nation who understand the science and know their future hangs in the balance. The tagline on RepublicEn's website spells out their stance pretty clearly: "We are the EcoRight. We want to solve climate change. We believe the free market is the answer." The website goes on to say, "Climate change is real. It is the duty of all Americans to reduce the risks. But to make a difference and take advantage of the opportunities, climate change needs to be addressed with free enterprise solutions instead of the status quo of ineffective subsidies and regulations."[36] Inglis himself says, "Conservatives can't embrace those solutions, but they can step forward with other solutions that fit with bedrock principles of free enterprise." The organization proposes limited government, accountability, free enterprise, and environmental stewardship as key principles for any climate policy.[37]

HOW WE FRAME THINGS MATTERS

Inglis and Dooley make the same point: it's not that all Republicans disregard science, but they all push back on what are perceived as liberal policies that will increase government control and decrease the choice of individuals and businesses. In 2019 when newly elected member of Congress Alexandria Ocasio-Cortez (D-NY) released a draft version of her and Senator Ed Markey's (D-MA) Green New Deal, conservative media and politicians bemoaned that it would take away people's cars, shut down air travel, and even make cows illegal. Sean Hannity of Fox News called it "government-forced veganism."[38] Even the *New York Times* editorial board, which supports parts of the plan, lamented that the Green New Deal also calls for measures for free higher education, universal health care, and affordable housing.[39]

What the reaction to the Green New Deal makes clear is that anything that comes from liberals that even hints at increasing government size and spending is going to get spun into some kind of socialist apocalypse ending the American way of life as we know it. Confirmation bias kicks in. If you are skeptical that climate change is a hoax created to prop up big government, then a policy like the Green New Deal reinforces the idea.

Bob Inglis and Debbie Dooley suggest that if Republican lawmakers can propose policy solutions to solve climate change with conservative, free enterprise-based approaches, they won't be as easy to dismiss.

In my opinion, there are lots of ways that we can figure out to decarbonize our economy. But in order to do so, we have to get both sides to the table. And in order to do that, we need to start speaking the same language. Liberals need to understand what's at the root of conservatives' fear of climate solutions. It may seem that given the short time frame within which we must solve the climate crisis, we don't have time to sit down and negotiate the conservative's concerns. But the truth is, we don't have enough time not to.

Bipartisan Efforts—the Climate Solutions Caucus

Bob Inglis was perhaps a few years ahead of his time. In 2016 Carlos Curbelo, a Republican congressman from South Florida, realized that something had to be done about climate change if the Florida Keys were going to stay above water. His constituents have been experiencing the reality of sea level rise and increasingly severe hurricanes for years. He reached across the aisle to his fellow Floridian, Democrat Ted Deutch. Together they formed the Climate Solutions Caucus—a bipartisan caucus to allow members of Congress to have an open dialogue about climate change and work out meaningful solutions that members of both parties can support. At the peak of its membership they had forty-five members from each party. Given that so few Republicans want to be on record as supporting climate policy due to the political liability of going against the party, forty-five is a pretty big number.

According to its official filing with the Committee on House Administration, "The Caucus will serve as an organization to educate members on economically-viable options to reduce climate risk and protect our nation's economy, security, infrastructure, agriculture, water supply and public safety."[40] However, as the caucus grew, it became a place for conversations to happen that otherwise wouldn't, rather than a place where policy was made. No substantial proposals came out of it, although in 2017 its members rallied against a spending amendment to halt the requirement that the military study climate change and publish its findings. The thirty-five Republicans who were in the caucus at the time got a few more colleagues to join in a vote to defeat the measure, and they easily won the fight.[41]

However, during the 2018 midterm elections, Curbelo and nearly half the Republican members of the caucus lost their seats.[42] As a result, the caucus has lost its claim to being a bipartisan force in the House, and it may not be very effective in its lopsided state at corralling Republicans to join together on key climate policy measures. Most environmental groups are not too sad about the caucus losing steam, as they thought the body wasn't effective to begin with. There were a number of instances in which members of the caucus could have been more vocal on important climate measures and weren't, and in which they could have taken action on key votes and didn't.

Still, in my opinion, Curbelo and his colleagues should be applauded for the effort to bring legislators together. The fact that they were able to amass a group of ninety members of the House of Representatives—a full 20 percent of the House, with a membership split equally between Republicans and Democrats—to publicly acknowledge their commitment to bipartisan solutions to climate change is a pretty impressive feat. And as with the efforts of John McCain and Bob Inglis, those of Carlos Curbelo and the Climate Solutions Caucus certainly help to pave the way for future bipartisan efforts to come. Already building off their success, in 2019 Senators Mike Braun (R-IN) and Chris Coons (D-DE) launched the Senate wing of the Climate Solutions Caucus, which includes Republican heavyweights Senators Lisa Murkowski, Mitt Romney, and Lindsey Graham.[43]

Young Republicans

Here's some more good news and an important trend to note if you're thinking of running for office or looking to keep your seat. Young Republicans recognize the threat of climate change and support the transition to clean energy much more than their elders. Surprised? I'm not.

Millennials have grown up understanding the reality of climate change and witnessing its effects; it is now common knowledge. They know that they are the ones who will be left holding the bag of a warming world. They know that Europe, China, India, and many other countries around the world are investing heavily in renewable energy. If we miss this opportunity, they also know that their future economic opportunities will be severely hampered. Who wants to be a member of the party that's clinging to dying polluting industries, dragging its feet as waves of innovation and opportunity pass it by?

In a recent poll of American views on climate change by the Pew Research Center, one of the most striking findings was the difference between millennial Republican views on climate and energy and those of Gen X and baby boomer Republicans. Among Democrats, millennial views were closer to those of the older generations. But Republican millennials had stark differences of opinion with their elders. More millennial Republicans than Republican Gen X and baby boomers believe in human-caused climate change. They are opposed to expanding fossil fuel usage, report witnessing the effects of climate change locally, and are supportive of renewable energy development, particularly solar (83 percent of millennials) and wind (87 percent).[44]

Thankfully, young conservatives aren't just standing on the sidelines. They're organizing, building coalitions, and making their voices heard. Michele Combs founded Young Conservatives for Energy Reform (YC4ER) after learning during her first pregnancy about the harmful effects of coal-fired power plants on nearby communities. She said, "I was a conservative, pro-family Republican all my life, and I thought, 'I can't believe we're not involved in this.'"[45]

Every year hundreds of young conservatives descend on Washington, DC, for their annual Conservative Clean Energy Summit with the hopes of letting their representatives hear their voices. These young conservatives support clean energy, and they want to see more of it.

They also know that climate change doesn't need to be a partisan issue. Their website features a telling quote from Ronald Reagan: "Preservation of our environment is not a liberal or conservative challenge, it's common sense."[46]

While partisan gridlock cannot be broken overnight, it can be overcome if influential voices are engaged in the policy discussion. In particular, conservatives who wield great influence among decision makers can be very effective. "We cannot transform America's energy future by only building support among voters and interests aligned with one party," this group of young Republicans astutely observes.[47] The goal, as they see it, is to "build an energy platform that speaks to the values of independence, security, prosperity, family, and stewardship."[48] Notice the framing. If a conservative Republican read this statement, how would their emotional Elephant brain react? These are words that resonate, that represent values they share with their community. This is exactly the type of messaging that won't spook their Elephant, making it possible for the message to connect with their Rider and potentially motivate people to action.

YC4ER aren't the only ones building platforms for young conservatives to advocate for clean energy and climate solutions while staying true to their values. Alexander Posner was a conservative twenty-two-year-old American history major at Yale University when he founded Students for Carbon Dividends, an alliance of thirty-four student organizations, conservative and liberal, from around the country—including twenty-three chapters of the College Republicans.[49] "I think a lot of young conservatives are frustrated by the false choice between no climate action and a big government regulatory scheme. They feel pressured that those are the only two options, and they're hungry for a conservative pathway forward on climate."[50] He added, "As young people with generations ahead we have the most to gain or lose from the issue."[51]

This bipartisan group of young adults is advocating for smart federal climate change legislation, and they actually have a specific policy in mind: the Carbon Dividends Plan, also known as the Baker-Shultz plan, named after former secretaries of state James A. Baker III and George P. Shultz. Their goal is to "catapult a market-driven climate solution—specifically the carbon dividends framework—into the national spotlight and open the door to a bipartisan climate breakthrough."[52] The carbon dividends framework they promote works like this: Producers of carbon dioxide emissions in the United States pay a carbon tax staring at $40 per ton and which gradually rises over time. But rather than the government keeping that money, it goes back to Americans as a dividend. Each family would get a check of about $2,000 a year.[53] Consumers who choose to use less carbon get to keep more of the dividend by paying less for clean energy sources.

Supporters of the framework include intellectual, business, and political leaders such as Ben Bernanke, former chairman of the Federal Reserve; Christine Todd Whitman, former governor of New Jersey and EPA administrator under George W. Bush; N. Gregory

Mankiw, former chair of the Council of Economic Advisors for George W. Bush; and, before he passed away, theoretical physicist Stephen Hawking.[54] However, the plan involves a troubling trade-off: to woo big business support and align with the conservative desire for small government, the plan restricts further government regulations on climate and cancels any liability the fossil fuel industry now faces for harm it has caused to date. (Frankly, I think both concessions are too big and won't get necessary traction, but perhaps there's room to negotiate.)

The carbon dividend framework has been promoted by the Climate Leadership Council, a policy institute whose founding members include ExxonMobil, ConocoPhillips, Shell, and BP. But fossil fuel companies are not the only companies that have hopped on this bandwagon. Johnson & Johnson, PepsiCo, AT&T, and Unilever are also on board. To round out this broad coalition, the council also features some of the largest environmental groups as founding members, including the World Wildlife Fund, the Nature Conservancy, and Conservation International.[55] Their strategic partners include the Worldwatch Institute, Bob Inglis's RepublicEn, and another carbon dividend advocacy group, the Citizen's Climate Lobby.[56]

So this plan has a lot of big names behind it. A majority of Americans also support it: a 2018 poll by Yale and George Mason universities found that 58 percent of registered voters support the idea of taxing carbon emissions with the revenue rebated directly to all Americans.[57] Surprisingly, even though the Green New Deal is much more progressive and includes a far larger role for government in its implementation, it's favored by even more Americans than the Carbon Dividends Plan. Another survey from Yale and George Mason shows overwhelming support for the Green New Deal, with 81 percent of registered voters saying they either "strongly support" (40 percent) or "somewhat support" (41 percent) this plan.[58]

Notice the silver lining here. Both conservative and progressive plans have support from the majority of Americans. The question, then, is how do we put it all together? Thankfully, the Green New Deal, while bold in its scope, is flexible when it comes to mechanisms to achieve that scope. A carbon tax and dividend approach could be a part of the mechanics of the Green New Deal. And thankfully, due to the popularity of the Green New Deal, Republicans are feeling the pressure to put forth more plans of their own. Proposals like the Carbon Dividends Plan that acknowledge the reality of climate change and are in line with conservative values.

Ryan Costello, a former Republican congressman from Pennsylvania, took over as managing director of the national bipartisan education and advocacy group Americans for Carbon Dividends after retiring from office in 2018. He described the situation this way: "You're seeing a lot of Republicans who want to engage and be part of a solution. . . . I think the politics of this are going to be, for many Republicans, is it sufficient to just look at what the far left want to do and call it crazy? Or is that going to be insufficient? Are they going to have to offer their own policy framework with which to address the issue?"[59] And thanks to the work of his organization and so many discussed in this chapter, there's now a growing body of conservative voices that together are helping to bolster climate and clean energy policy in America.

CHAPTER 5

JOBS, JOBS, JOBS

Every revolution was first a thought in one man's mind, and when the same thought occurs to another man, it is the key to that era.

—RALPH WALDO EMERSON[1]

IN THE PREVIOUS CHAPTER I talked about how different policy mechanisms could help us transition to a cleaner, greener economy. And while I certainly hope that the federal government starts putting its back into the renewable energy transition, frankly the transition is already happening, despite the fossil fuel industry's best efforts to slow it down and despite the lack of policy support at the national level.

Why? One word: cost.

In April 2019, for the first time in US history, more electricity was generated by renewable energy than by coal.[2] Think about that for a moment. Coal is the incumbent technology; it's been the dominant source of energy in this country for over two hundred years. Not to mention that a coal lobbyist, Andrew Wheeler, was heading up the Environmental Protection Agency at the time, that Trump pulled the US out of the Paris Agreement, and that Obama's Clean Power Plan stalled in courts, never to see the light of day. And yet somehow renewable energy has overtaken coal. How can this be? The answer is that *market forces* are putting coal plants out of business while making renewable energy the fastest growing technology in the energy sector.

DECLINING COSTS OF TECHNOLOGY

What makes renewable energy the inevitable dominant source of future energy is that it's, well, renewable. Meaning the source of the energy, like sunshine and wind, are free. The primary cost is the technology needed to capture the energy: the solar panel or the wind turbine. And as we all know very well, technology gets better and cheaper as more customers adopt it.

Take computers, for example. In the 1960s an IBM computer "cost as much as $9 million and required an air-conditioned quarter-acre of space and a staff of 60 people" to run.[3] Now a fifth grader with a smartphone has access to literally millions of times more computing power than what NASA used to send astronauts to the moon.[4]

How does this progression work? Technology cost curves often follow a similar pattern. As a technology catches on and sales go up, the producers can scale production and invest in research and development, allowing them to benefit from economies of scale and technological innovation, creating cheaper per-unit costs. Competition among producers creates more downward pressure on the price of the technology, which benefits the consumer and makes the products more attractive to a larger audience. As the market grows, technological breakthroughs; streamlined manufacturing and distribution; and public familiarity with the technology create a positive reinforcement loop. In other words, as the costs come down, more people buy it, which means more people see it, which means more people buy it, which means the costs come down further, and the cycle continues.

For solar energy in particular, this trend has been at play since the first modern solar panel was invented in 1954 in Cherry Hill, New Jersey, at Bell Laboratories. In fact, the cost drops for solar technology have been so predictable that Richard Swanson, founder of SunPower, an American solar panel manufacturer, theorized the law of declining solar costs, which is now known as Swanson's Law.

Essentially it says that each time solar deployment doubles, the cost will drop by 20 percent.[5] And what we've seen is that such a drop in price over time is dramatic: in the past ten years alone, the costs of solar panels have dropped 70 percent.[6] And the price of solar is continuing to fall. According to consulting firm Wood Mackenzie, the price of photovoltaic modules—assembled groups of solar cells—is likely to fall 40 percent from 2018 to 2022, from $0.30 per watt-DC to $0.18 per watt-DC.[7] This disruptive technology not only has a limitless free fuel source, with the costs of the equipment continuing to decline, but it's also a source of zero emissions clean electricity.

Now let's look at the incumbent technology. Fossil fuels have the opposite economic dynamics. The technology is cheap because it was developed a long time ago with heavy subsidies from the government, which continue to this day. But the sources of this fuel are limited, and the resulting emissions are dirty. Since the world has been burning through its more easily accessible reserves of oil, gas, and coal, for over two hundred years, they're becoming harder to find and more difficult, destructive, and expensive to extract, refine, and ship to customers. Shale oil, tar sands, fracking, deep sea drilling, and mountaintop removal are examples of the extreme extraction practices and dirty fuel sources the fossil fuel industry is now forced to pursue in order to keep costs low. But these strategies will only keep fossil fuels competitive temporarily. As more people than ever before need increasing amounts of energy, and as the supply of easily accessible fossil fuels continues to dwindle, the price will keep going up, making it less and less competitive with renewables.

For example, a 2019 report by the Rocky Mountain Institute, a nonprofit research group focused on sustainability, predicts that by 2035 the cost to operate a natural gas plant in America will exceed that of building new wind and solar infrastructure. This means that the natural gas plants being built today will be retired years before they've even been paid off. In other words, fossil fuel–stranded assets will become a massive financial liability in the coming years. In fact, in May 2020, *Bloomberg* reported, "Onshore wind and solar PV are now the lowest-cost new sources of power generation for at least two-thirds of the global population, 71% of global GDP, and 85% of global power generation."[8]

What many don't realize is that we're living through a historic moment. For decades, the cost of fossil fuels has been rising, and the costs of renewable energy have been declining. We're now at that critical juncture where the costs of fossil fuels will increasingly become more expensive than renewables, and renewables will become increasingly cheaper than fossil fuels. For those holding out for nuclear energy, the length of time it takes to build nuclear power plants, the costs of building and insuring them, as well as the costs of finding, extracting, processing, and disposing of uranium safely and without proliferation makes this option simply noncompetitive from a cost perspective.[9] Even though every so often you hear of well-meaning folks making an argument for next-generation nuclear, the economics speak for themselves, and investors have lost their appetite for the technology.

If you're looking to catch hold of the next big trend, whether that's from a career perspective, an investment strategy, or concern for life on Earth, pay attention to what this all means. We are moving into a world in which renewable energy will be the cheapest and most abundant source of energy worldwide. The fossil fuel industry will go the way of so many now-defunct technologies that were made obsolete by innovation.

And it's already happening, with or without federal policy. According to consulting firm McKinsey and Company, the share of the global energy mix represented by renewable energy will increase from 25 percent today to over 50 percent by 2035 and nearly 75 percent by 2050.[10] Let that sink in for a second. Can you think of an economic opportunity as big as transforming global energy production in such a short time frame? (See figure 5.1.)

Here in the US, total installed solar photovoltaic (PV) capacity is expected to more than double from 2019 to 2024.[11] To put that in perspective, it took the US over forty years to install sixty-five gigawatts of solar. We're going to double that number in five years. Noticing a trend here?

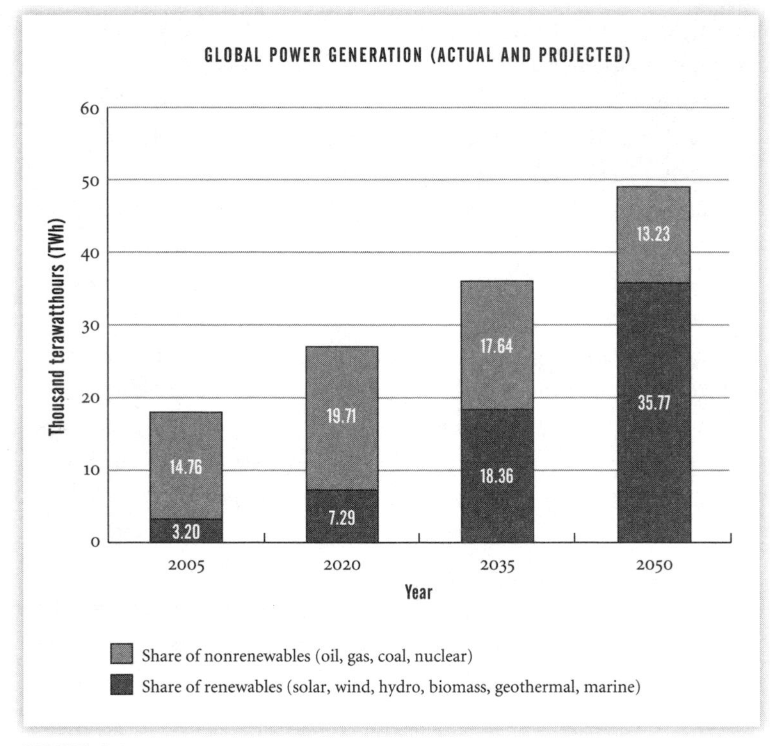

FIGURE 5.1

100% CLEAN ENERGY

Mark Jacobson, professor and researcher at Stanford University, thinks we can do even better than the current projections. He and his team have developed a state-by-state model for how the US can reach 85 percent renewable (specifically wind, water, and solar) by 2030, and 100 percent by 2050.[12] Here's how the 100 percent clean energy mix would look: On and offshore wind would cover about 50 percent, and solar around 45 percent. The other 5 percent would come from hydro, geothermal, wave and tidal. Jacobson calls this mix Wind, Water, and Solar (WWS). Note, this doesn't include any nuclear power, carbon capture and sequestration, or biofuels. The key to his plan is that overall we'll reduce the amount of energy we need by close to 40 percent by "electrifying everything." By converting combustion motors to electric motors, we lose a lot less energy to heat and thus increase the energy-to-work conversion.[13]

Jacobson agrees that this plan is ambitious and would require strong political will. It helps that his plan is projected to cost 50 percent less than business as usual.[14] That's right, folks. Pollution-free energy that'll save consumers $1 trillion a year on energy costs.

It also produces a net gain of two million full-time jobs.[15] Remind me how this is still a debate?

If you don't believe the experts at Stanford University and McKinsey and Company, what's the fossil fuel industry's perspective on future energy scenarios? According to Shell, one of the world's largest oil and petrochemical companies, solar will surpass oil becoming the world's largest energy source by as early as 2050.[16]

While predictions about the energy future timeline vary, the overarching story remains the same. With or without the federal government's support, we're quickly moving to a renewable energy–powered country and world strictly because it's cheaper to produce. That means new technologies, new industries, new manufacturing processes, and new business models will reshape our society in the coming decades more than we can possibly imagine. And it's widely accepted that renewable energy employs more people per unit of energy created than fossil fuels.

Long story short, that means new jobs, and lots of them. Manufacturing jobs, sales jobs, installation jobs, maintenance jobs, customer service jobs, the majority of which can't be outsourced. But this isn't just a pipe dream. We're already seeing the growth of jobs in the renewables sector outpace the rest of the economy dramatically.

JOBS

According to the 2019 *Clean Jobs America* report released by the nonpartisan business group E2 (Environmental Entrepreneurs), clean energy jobs already outnumber fossil fuels jobs nearly three to one (3.26 million to 1.15 million).[17] Keep in mind that renewable energy only makes up 11 percent of the total energy mix now.[18] Think about how many jobs that means when renewables become the dominant source of energy over the coming decades.

In 2018, the fastest-growing jobs in twelve different states were in renewable energy. The US Bureau of Labor Statistics predicts that from 2018 to 2028 we will need many more wind turbine technicians and solar installers to meet the growing demand across the country.[19] Solar panel installers in particular will be in high demand: the Bureau of Labor Statistics predicts that the number of solar panel installers in the US in 2019, about 9,700, will swell by 63 percent, to 15,800, over the following ten years.[20]

Another way of quantifying job growth in renewables is to note that jobs in solar and wind energy are currently growing twelve times faster than the rest of the economy.[21] In the electricity sector in particular, solar is by far the largest job creator. In 2016 solar power accounted for 43 percent of the power sector's workforce, while coal, oil, and gas combined accounted for only 22 percent, according to the Department of Energy.[22] (See figure 5.2.[23])

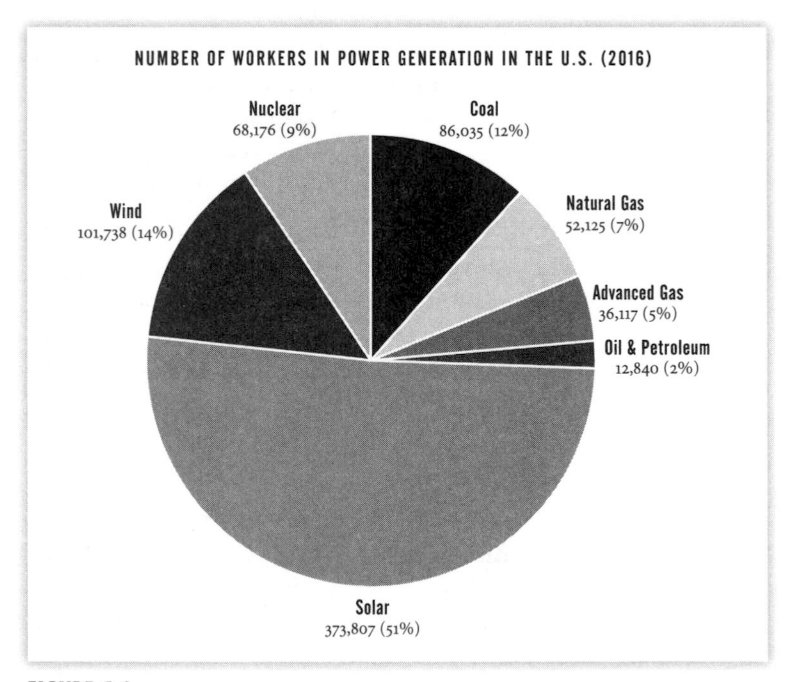

FIGURE 5.2

Keep in mind as you look at the graph on the previous page—solar accounts for less than 2 percent of the electricity generated in the country. Can anyone still credibly claim that renewables are bad for jobs?

WHO BENEFITS?

Some people ask where the economic benefit of renewable energy will go. Will it only benefit liberal enclaves, they wonder, where strong policy has supported clean energy growth? The answer is far from it. Red states are by far the biggest producers of renewable energy in the country. Of the top ten solar and wind producing states by percentage of in-state electricity generation only two (California and Vermont) are historically blue states. (See figures 5.3 and 5.4.[24])

Now that we know how big an opportunity clean energy is, let's hear from a few people who are working hard to make sure that it creates real benefits for American communities.

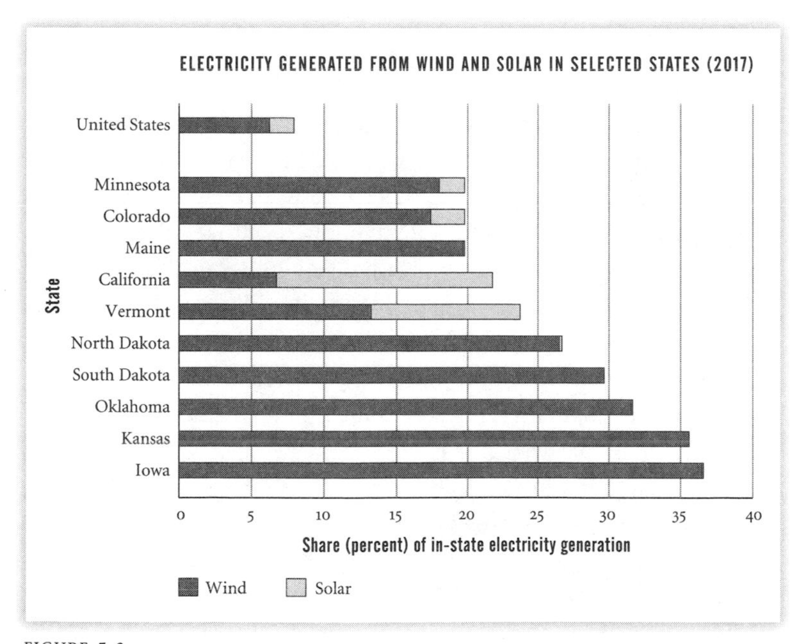

FIGURE 5.3

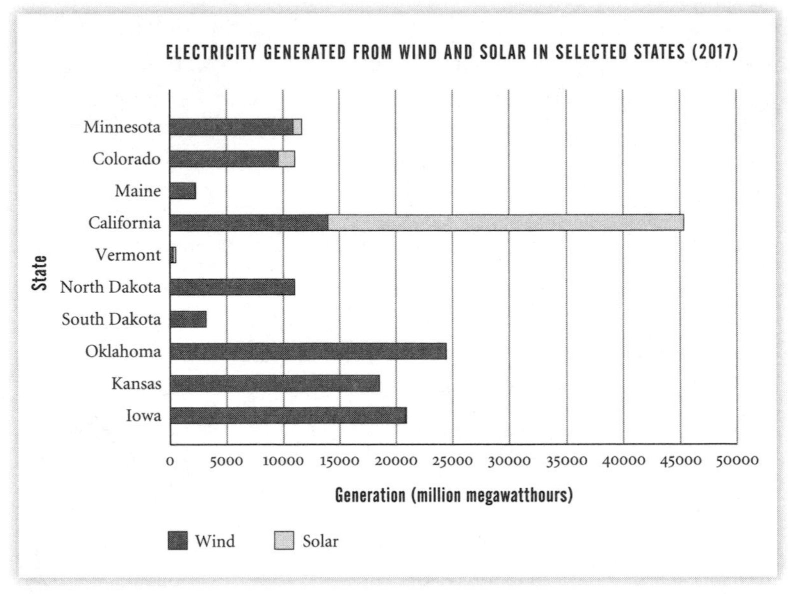

FIGURE 5.4

Green for All

CNN host Van Jones began his career as a lawyer and social justice activist, and in 2008 he started a nonprofit called Green for All. Its mission is to build a "green collar economy" strong enough to lift people out of poverty. On the one hand, he saw that African American communities had disproportionately high levels of unemployment and needed pathways out of poverty. He also saw that in order to ameliorate climate change, we would need to create millions of new jobs retrofitting communities. These were jobs that couldn't be outsourced, like installing solar panels and weatherizing homes. Jones's idea is that we can solve the climate crisis and put people who need a job back to work at the same time.[25]

Since the 1980s, as neoliberal economic policies allowed US companies to ship their manufacturing jobs overseas, working class Americans have been struggling to find good jobs that will support their families. As Jones so articulately puts it, "This is the chance for

America to finally return to its roots, as the most important economy in the world, not because we are the number one consumers, but because we are the number one producers."[26]

Green for All was one of the first national organizations to connect issues of race, economics, and environment and to highlight the interrelated nature of these issues to the public.[27] The four pillars of their work include working at the federal level, where they publish white papers and advocate for Green Jobs legislation; working at the local level with mayors to develop green economic development plans; working at the personal and community levels through their Green for All Academy, which trains young leaders to become champions for an inclusive green economy; and working with the business sector, including connecting green entrepreneurs to capital.

Jones's book *The Green Collar Economy* was the first environmental book written by an African American to make the *New York Times* bestseller list. Jones quickly rose to popularity in the national spotlight as a brilliant orator with the ability to connect issues and predict economic trends. Eventually this led to his appointment as special advisor for Green Jobs, Enterprise, and Innovation at the White House Council on Environmental Quality, for President Obama.[28]

In 2008 Jones was off to a great start in the Obama White House, crafting initiatives that would become key to Obama's economic and environmental strategy. But then politics got involved. Conservative media outlets started digging up his history as a community organizer who for years had worked tirelessly for racial equality and criminal justice reform. They ran a vicious smear campaign, painting him as radical for the purpose of tarnishing Obama's reputation. Unfortunately, Jones decided to step down rather than let it affect the work that Obama was trying to accomplish. "I cannot in good conscience ask my colleagues to expend precious time and energy defending or explaining my past. We need all hands on deck, fighting for the future," Jones wrote in his resignation letter.[29]

With the departure of Van Jones from the White House in 2009 and the financial crisis in full swing, Obama's plans to invest in building the green collar economy were put on hold. It's sad to think about the decade that was lost in the climate fight because of this round of political sparring. And the previous rounds that lost us the two decades before that. But perhaps there's a lesson here.

Putting all our efforts into electing the right climate candidate is not enough. The fickle nature of politics means that any win, big or small, is subject to change at any moment. That's all the more reason to find the climate courage to work in our local communities to make change happen. Starting from the ground up is the only way to ensure that new sustainable ways of living are woven into the fabric of our culture. And thanks to the economics being increasingly in our favor, that's exactly what's happening.

Coal Miners to Solar Installers

Dan Conant grew up in West Virginia. After college, he worked in the solar industry in Vermont, got a master's degree in environmental science and policy, and moved to Washington, DC, to work for the US Department of Energy. When he saw the growth potential of solar, he knew that it could help bring jobs back to his home state.[30] West Virginia, like so many places, needed something new. The coal industry had been dying for years. Not because of regulations but because cheap natural gas, renewable energy, and energy efficiency have made old polluting coal-fired power plants uneconomical. Since West Virginia's main form of employment for generations has been coal mining, people have been slowly losing jobs for years. It's not a good situation. Without much economic opportunity, cycles of poverty perpetuate, which contributes to the state's growing opioid epidemic, one of the worst in the country.

With his experience in the industry and his entrepreneurial spirit, Dan Conant headed back to his hometown of Shepherdstown, West Virginia, where he started the business Solar Holler in 2013. Since then, he has been successfully installing solar for nonprofits, community centers, and low-income residents. But Conant wasn't satisfied only hiring qualified workers to meet his business needs—in fact, he realized that there weren't enough qualified workers to keep up with demand. He also wanted to create sustainable jobs for those who really needed them but who didn't yet have the necessary skills.[31]

With these people in mind, Dan partnered with fellow native West Virginian Brandon Dennison to create a truly unique partnership. While Conant was getting Solar Holler off the ground, Dennison had been working on the flipside of the coin, starting an organization called Coalfield Development in 2010 that trains former coal miners to work in social enterprises that benefit the community.[32] Trainees from Coalfield Development started working at Solar Holler, where they do a two-and-half-year apprenticeship learning the technical skills and gaining the job experience needed to continue with careers in solar energy. Solar Holler's is the first solar job training and apprenticeship program in West Virginia.[33] But hopefully not the last. It's also critical that around the country we learn from Solar Holler's experience, especially in places where fossil fuel workers will need new skills and new employment opportunities as our economy transitions to clean energy.

Solar Richmond

Richmond, California, is home of the infamous Chevron oil refinery. For over one hundred years, this refinery on the northeast side of San Francisco Bay has processed 250,000 barrels of oil a day. The oil comes in by ship and leaves as processed gasoline, jet fuel, and other chemical products to be sold around the country and world. The refinery creates some of the worst pollution in the country. Richmond, like so many communities where dirty industries site their production facilities and dumping grounds for toxins, is predominately populated by people of color, with about 40 percent Hispanic and 25 percent African American residents.[34] And like so many communities of color that are the victims of direct pollution in their backyards, the health impacts on the city's residents are dumbfounding. In 2012, the asthma rates for residents who had lived in Richmond for fifteen years or more were an astounding 45 percent.[35] In addition, they have extremely high rates of cancer and pulmonary disease, as well as acute health problems such as eye irritation, headaches, and nosebleeds that over 63 percent of residents report experiencing regularly.[36]

Economically, the city has had some real challenges too. While things have been getting better over the past ten years, back in 2009, Richmond had an 18 percent unemployment rate and the third highest crime per capita rate in California.[37]

A few years prior, in 2006, tech industry veteran Michele McGeoy saw that there was a way to help people in Richmond turn things around: the green economy. Michele worked with the city of Richmond and other community leaders in the area to put together a first-of-its-kind solar workforce training program for underemployed residents of Richmond looking for a new opportunity. They called it Solar Richmond. Solar Richmond started by training local residents in solar installation, but it has continued to evolve, and now its programs include leadership development and job placement through a referral program. Solar Richmond is now considered a national model.

Shalini Kantayya is a documentary filmmaker who, in 2016, released a film called *Catching the Sun*, which featured Solar Richmond. She made the movie to tell a different kind of climate story—one filled with accounts of climate courage. "I feel really disenchanted with the traditional climate movement and its messaging," she told *Grist*. "I really think that there's a story that's being lost in these doom and gloom climate scenarios, which is the story of hope. When we're talking about what motivates people to effect change, I think it's actually hope."[38] I couldn't agree more. In fact, what we've learned so far throughout this book supports her sentiment completely.

"When I first visited Solar Richmond," Kantayya said, "I was moved by this small solar training program set in the backdrop of this old oil refinery town. Here's a program placing people who are at the heart of the crisis into the heart of solutions. I saw Solar Richmond as a microcosm of the kind of transformation that cities across the United States need to make if we're going to meet the challenge of climate change."[39]

What's important to note here is that models like these are not just important for the local community they serve. They're important demonstrations of what's possible when people come together to try and solve problems that are affecting their community. The innovative ideas that are pioneered in one community can be replicated in communities everywhere. Imagine if we had Solar Richmond training programs around the country.

Of course, every climate courageous organization trying to create change faces its obstacles and challenges. In a recent conversation with Cheryl Vaughn, executive director of Solar Richmond, I learned that the organization is going through a tough period. When the organization first started, there was a strong partnership with the City of Richmond. The partnership has fizzled as certain staff from both the city and Solar Richmond have moved on. Their attempts to establish strong job pipelines for their trainees with solar companies in the area hasn't panned out as expected. When the organization started, it was the first of its kind and received lots of support and funding. Now fundraising efforts have slowed down.

As every entrepreneur knows, the life of an organization goes through many highs and lows. Given the explosive need for solar installers in the coming decade, Solar Richmond's unique services will continue to be in strong demand. As a movement that needs to make great strides quickly, we need heroes like Cheryl Vaughn to keep fighting and for people to keep building programs like Solar Richmond, even though success is not always guaranteed. It's only through trying that we learn what works and only with that knowledge can we improve our collective strategies.

GRID Alternatives

In Oakland, California, just south of Richmond, is another green job training powerhouse, GRID Alternatives. GRID is the nation's largest nonprofit solar installer: it has installed solar panels for over fifteen thousand low-income family households in the US and is increasingly serving communities around the world. More impressive, though, is that they've trained and educated over forty thousand people in solar installation and other aspects of the industry in the classroom and in the field. These trainees now have hands-on experience on the roof and are able to translate that experience into job opportunities.[40]

GRID has been a pioneer in the field in many ways. They're not just training installers. They also have classroom educational programs for students from kindergarten all the way through high school. Their spring break program gives college students the ability to get in-depth solar installation training. They also have a fellowship program for AmeriCorps volunteers who join the GRID team for a year to help deploy renewable energy and implement job training in underserved communities while gaining hands-on training and technical skills themselves. GRID also offers training programs specifically to help women and veterans break into the industry.[41] I'll talk more about GRID in chapter 8.

Mark Jacobson's research at Stanford makes clear that a lot of people currently employed in the fossil fuel industries are going to lose their jobs as the world transitions to new clean energy technologies and the fossil fuel industries become a thing of the past. However, we also learned from his projections that the new clean energy economy has the potential to provide a net increase of two million jobs. In order to ensure a smooth and just transition into the clean energy era we know is coming, we have to come up with a realistic way to train the millions of people who will be transitioning out of jobs in the old energy economy and create pathways for meaningful employment in the new energy economy.

Despite the success of the training programs I'm highlighting, they're not going to be sufficient to do the job alone. Far from it. But they can serve as models that can be replicated and scaled up in every community in the country. The transition to sustainability is going to be riddled with questions about best practices and challenges implementing even the best plans. To navigate the necessary changes rapidly is going to require experimentation, creativity, and entrepreneurial spirit at the start. Then it's going to require scaling up the solutions that work. These examples demonstrate what's possible and highlight the courageous spirit of the innovators who are exploring the unknown so that the rest of us can learn from their journeys.

CLEAN ENERGY MANUFACTURING

As important as rooftop solar jobs are, they are just one part of the clean energy economy. Let's not forget another major opportunity: manufacturing. How many industries are bringing manufacturing back to the United States these days? Not many. But one is: clean energy. One innovator helping to bring manufacturing back to the US? Elon Musk.

One of the many companies the billionaire entrepreneur has started is the world-renowned electric vehicle company Tesla. While Musk was working on building the world a better vehicle, his cousins Peter and Lyndon Rive were equipping it with solar panels through their company SolarCity, which Elon had helped cofound and where he served as chairman of the board. In 2016 Elon spearheaded merging the two companies, and Tesla bought SolarCity, putting the two under one roof.[42] Why? Tesla's electric cars need state-of-the art batteries to keep those puppies cruising. As a result of Tesla's relentless drive to improve battery storage for its cars, SolarCity's residential and commercial solar customers will increasingly benefit from improved batteries as well, not to mention the opportunities for developing and producing utility-scale batteries, those big enough for an electric utility to use.

Battery storage has long been considered the holy grail of clean energy. With efficient storage, you can capture all the sun's rays during the day and use them to power your home at night, or capture the wind blowing at night and use it to power your community during the day. Thanks to the efforts of companies like Tesla that are developing improved batteries and ramping up production, battery capacity continues to increase, and the costs of clean energy storage continue to come down. According to the Department of Energy's National Renewable Energy Laboratory, utility-scale battery storage available today that can hold a charge for four hours, six hours, and eight hours is now cost competitive with traditional fossil fuel–powered peaker plants (typically natural gas plants that utilities turn on when demand for electricity is highest). Long story short, battery capacity is already here for utilities and is becoming increasingly economical for homeowners and businesses to store energy too.[43]

Of course, speaking of Tesla, one of the biggest opportunities to combat climate change is to rapidly electrify the transportation sector and power it with clean energy. The trifecta that Tesla is aiming for is to provide each American home with a Tesla solar roof, that charges a Tesla Powerwall battery bank in the garage, which charges your Tesla vehicle overnight. And to see that vision come to light, Musk has brought American manufacturing expertise to bear. The Tesla automobile factory, based in Fremont, California, employs ten thousand people, making it one of the largest manufacturing employers in the state.[44]

In Reno, Nevada, Tesla has also built what's known as the Gigafactory to assemble batteries for its cars and for its Powerwall home battery packs, as well for utility-scale battery storage. With a footprint of 5.5 million square feet, the Gigafactory is literally the largest building in the world—one hundred Boeing 747 jets could fit inside it.[45] The Gigafactory is projected to create over $100 billion in economic benefit over two decades from the construction and operation of this factory, according to Nevada governor Brian Sandoval.[46]

Hop over to Buffalo, New York, where Gigafactory 2 is building the solar component of Musk's technology suite. Musk has decided to rethink how solar panels are designed. Rather than manufacturing solar panels that attach to your roof, he wants to build the solar cells right into the roofing materials—the Tesla Solar Shingle. Of course, true to all of Musk's products, the style and elegance of the product match its engineering. The Buffalo facility exceeded its 2020 staffing targets of 1,460 employees in Buffalo, including outside the Gigafactory, and was employing 1,800 by February 2020.[47] The final goal is five thousand jobs to New York employees within a decade of the facility's completion.[48]

Naturally, there have been hiccups in Tesla's progress. In 2018, while hitting its target for delivering cars, it also had to lay off 7 percent of the company staff. In addition, the residential solar sales had a slight decline from 2017 to 2019. While some state officials in New York had expressed concerns about Tesla's ability to hit their job projections in 2020 and beyond, the good news is, if they don't hit their numbers, they'll owe the state of New York $41.2 million each year their projections are off.[49] More recently however, the company had a very strong and profitable second half of 2019, resulting in a record-high share price and company valuation in February 2020.[50]

No one can predict what the future will look like for Tesla, although many people do try. What I can say is this: in the beginning of every industry, there are a lot of hurdles to overcome. If we look at the internet, for example, in the late '90s and early 2000s there was a tremendous explosion of internet start-ups. Investment dollars flowed into them. Then the dot-com bubble burst, and a good number of these start-ups came crashing down. However, only a few years later, companies such as Google, Facebook, and Amazon started to emerge as global powerhouses.

Despite the challenges Tesla is facing, this clean energy and electric vehicle company has built or refurbished three new American factories, some of the biggest in the world, producing cars, solar panels, and batteries. Not only are these technologies good for the planet, but by manufacturing them here in the US, Tesla is creating jobs for Americans and creating real economic value for American communities.

Buffalo, like Cleveland, Pittsburgh, and Detroit, was an iconic, bustling American manufacturing city during much of the twentieth century. It was a leader in everything from shipbuilding to car manufacturing to steel manufacturing. Republic Steel and Bethlehem Steel's Lackawanna plant were major employers in the city.[51] In fact, the Gigafactory 2 sits on the very site where Republic Steel operated a mill from the early twentieth century until it closed its doors in 1984.[52]

Could Musk be starting a new manufacturing trend? Will other manufacturers follow his lead and bring manufacturing to places like Reno and Buffalo where people are looking for work? Let's hope so. "We actually did the calculations to figure out what it would take to transition the whole world to sustainable energy. You'd need 100 Gigafactories," quips Musk.[53] A hundred new Gigafactories in the US producing items like solar panels, batteries, and electric vehicles could be just the ticket, not only to bring needed jobs and economic activity back to our local communities but also to confirm the US as a leader in the global transition to a sustainable economy.

In fact, it looks like the trend is already catching on. In January 2020, GM announced it would invest $2.2 billion to upgrade its Detroit-Hamtramck assembly plant, to become its first plant dedicated solely to manufacturing electric vehicles.[54] The plant, set to come online in 2021, will manufacture all electric pick-up trucks, SUVs, and even the electric Hummer. Yes, you read that correctly: Detroit will soon be pumping out electric Hummers that go 0 to 60 mph in three seconds, boasting "1,000 horsepower and 11,500 pound-feet of torque."[55] Feeling climate courageous yet? By the way, in the process GM will create 2,200 good-paying manufacturing jobs. And the batteries for these vehicles will be manufactured in a new $2.3 billion facility built by GM and LG Chem in Lordstown, Ohio, which will create an additional 1,100 jobs.[56]

CHAPTER 6

FAITH COMMUNITIES IN ACTION

In every outthrust headland, in every curving beach,
in every grain of sand there is the story of the earth.

—RACHEL CARSON[1]

FOR BETTER OR WORSE, the religions of the world have, for much of human history, played a critical role in determining our collective sense of right and wrong. While they differ in their approaches to many things, one thing that most faiths share is a sense of respect for nature. Whether it's called caring for God's Creation, being a steward of nature, honoring Mother Earth, Gaia, or Pacha Mama, there is a common thread across spiritual traditions that recognizes the sacredness of the natural world.

Here in America, the modern environmental movement began with people of faith. Ralph Waldo Emerson, Henry David Thoreau, and John Muir, among many others, based their conservation efforts on a spiritual ethos. To them, nature was a place to commune with the divine. "The clearest way into the Universe is through a forest wilderness," wrote John Muir, a pioneer of climate courage upon whose shoulders the entire environmental movement stands. "Keep close to Nature's heart," Muir said, "and break clear away, once in a while, and climb a mountain or spend a week in the woods. Wash your spirit clean."[2]

However, these naturalists witnessed the American landscape being threatened by rapid industrialization during the nineteenth century. Their writings helped create an appreciation for nature in the public eye, and their activism led to the start of environmental advocacy groups, such as the Sierra Club, founded by John Muir in 1892, and to the eventual founding of the national parks system.

Today the shared conservation ethos of former generations has sometimes been lost in our polarized politics. But is it true, as we hear in the news, that faith communities in America have discounted the evidence for climate change? Does the belief in a higher power preclude one from working to stop it? Does one's acceptance of climate science make someone a godless heathen? Well, it depends. For religious groups that want to remain loyal to the Republican Party, climate change has been dismissed as a hoax perpetrated by liberals—and in particular liberal scientists, who don't share their faith. For groups like these, science has long been viewed as an attack against their faith, and climate change is simply the latest wave.

However, there's another story that doesn't get as much airtime. A story left out of the climate change narrative. Many faith groups around the country and the world have become engaged on climate change in a way that we haven't seen since the civil rights movement. These groups are marching in the streets, organizing in their communities, voting at the polling booths, divesting from fossil fuels, and investing in clean energy. Here are some of the leaders and their bold initiatives to engage people of faith in the fight for the planet.

THE POPE'S ENCYCLICAL

In 2015 Pope Francis released a papal encyclical—an urgent message addressed to all Catholics—entitled *Laudato Si'* [Praise Be to You]*: On Care for Our Common Home*. It caught the attention of people around the world because in it, the pope decried climate change, especially its effects on the poor. It was quickly dubbed *The Encyclical on Climate Change and Inequality*.

A papal encyclical is rare and therefore significant. Encyclicals are written when a pope feels a pressing need to clarify the Catholic Church's teachings on a given subject: "They deal with complex social and moral issues and back up their claims with reference to the Bible and to Catholic tradition and doctrines."[3] *Laudato Si'* was the first encyclical that directly addressed the environment and sustainability. The message Pope Francis sent to the 1.2 billion Catholics around the world by way of the *Laudato Si'* is loud and clear: climate change is real, caused by humans, and we have to do something about it. But the encyclical also makes clear how one's identity as a Christian relates to climate change: "The ecological crisis is also a summons to profound interior conversion. . . . Living our vocation to be protectors of God's handiwork is essential to a life of virtue; it is not an optional or a secondary aspect of our Christian experience."[4]

What prompted this strong stance from the pope? To understand, it helps to have a sense of Pope Francis's story. Born Jorge Mario Bergoglio in 1936, Pope Francis grew up in Argentina during tumultuous times. He saw firsthand the ills of inequality in a country where the wealthy lived in a bubble while many others lived in abject poverty. Jesuit priests, in the tradition of Saint Francis, are tasked with going to the frontier of poverty. Pope Francis, throughout his priesthood, would spend his time in the most impoverished neighborhoods of Buenos Aires serving and ministering to the poor. As a bishop, he never had a car, and he wouldn't accept a ride. He always wanted to walk the streets or to take the subway or the bus. In the neighborhoods and on public transportation he talked with people and learned about their struggles. They referred to him as the Priest of the Villas and the Priest of the Slums. Even now as the pope, he chooses not to live in the papal residence but rather in the guest house, so he can live a more humble, simple life.[5]

It's no wonder, to me at least, that this pope, who is known for walking the streets and washing the feet of the poor, has used his seat of authority to call the world's attention to climate change, which will disproportionately affect the poor. The pope describes the correlation this way in the encyclical:

> We are faced not with two separate crises, one environmental and the other social, but rather with one complex crisis which is

> both social and environmental. Strategies for a solution demand an integrated approach to combating poverty, restoring dignity to the excluded and at the same time protecting nature.[6]

One doesn't have to be Catholic to find truth in that statement, nor in this:

> If we approach nature and the environment without . . . openness to awe and wonder, if we no longer speak the language of fraternity and beauty in our relationship with the world, our attitude will be that of masters, consumers, ruthless exploiters, unable to set limits on their immediate needs. By contrast, if we feel intimately united with all that exists, then sobriety and care will well up spontaneously.[7]

Wow.

I remember when the encyclical came out. It was free online, and I printed it out and carried it around with me, reading from it every chance I could get. A few weeks later, I had a meeting with former head of the EPA under George H. W. Bush, Bill Reilly. Bill is a fascinating guy. He's a staunch Republican and yet a passionate environmentalist. It was largely because of him that President Bush supported environmental efforts like the Montreal Protocol, which limited ozone-depleting chemicals, and signed the United Nations Framework Convention on Climate Change at the Rio Earth Summit in 1992.

When I met with him, Bill had just come back from meeting with the pope. He showed me pictures and told a few stories. Bill was of the belief that the encyclical would have a tremendous impact on shifting Republican views on climate change. He told me to start carrying around a copy of it. I took out the copy I had in my bag, and we had a laugh.

Pope Francis used his pulpit to call to the attention of his spiritual followers the serious threat of climate change. Thankfully, he wasn't the first. In the US, there has been a growing movement of religious leaders organizing congregations to do the same for about two decades, including a visionary priest from San Francisco, my friend the Reverend Canon Sally Bingham.

INTERFAITH POWER AND LIGHT

Sally Bingham's road to becoming an ordained priest, let alone a pioneer in the field of ecology and religion, was not one she had ever planned on. Nor was it an easy road for her. In the early 1980s Sally moved with her husband and three children from New York to San Francisco. She enjoyed her life as a housewife and a stay-at-home mom. She regularly attended her local Episcopalian church on Sundays with her family. And she also served on the board of the nonprofit Environmental Defense Fund.

As someone who grew up in Woodside, California, and spent most of her childhood outside, she had always had a deep love for nature. But it was during her time serving on the EDF board that she became alarmed by climate change, pollution, and what humans were doing to the planet. As she learned more about the state of the environment, she noticed that no one at church was talking about it. They were praying for the earth, but they were also driving SUVs, living consumerist lifestyles, and polluting their neighbors' air and water. "We were destroying this thing we are supposed to have reverence for," she told me.[8] She felt they were unknowingly being hypocrites, if not sinners.

She started speaking to the church leadership about the issue. If we are taught to love our neighbor, she asked, why are we ignoring the damage we're doing to them and the environment through our behavior? After many conversations and no easy answers, a bishop suggested that she consider going to seminary to explore the question more deeply, study the theology behind it, and try to understand the source of the disconnect between the church's stated beliefs and what they preached from the pulpit, in terms of congregants taking action and changing their behaviors. And that's exactly what she did.

But Sally had never gone to college, and she needed an undergraduate degree before she could attend the seminary. At the age of forty-five, with one of her children in college, Sally signed up for a four-year undergraduate degree at the University of San Francisco. After four years of college, four years of seminary, and another two years of serving in a supportive role in a local church, Sally became an ordained priest at the age of fifty-five.

During this time of great transformation, she also had to endure personal challenges. Her marriage ended in divorce. She had no job and no financial support, but she persisted in her mission nonetheless. "My own motivation comes from saying 'yes' more often than not. Yes to God and yes to stretching my imagination," she writes in her book *Love God, Heal Earth*.[9]

For the next twenty years Sally would go on to build one of the first and largest organizations aimed at organizing different religious faiths around their care for the earth, Interfaith Power and Light. Before we continue with her story though, stop for a moment and think about the courage—the climate courage—it took, at the age of forty-five to give up the life she knew to enroll in college, continue to seminary, and become a priest, in order to follow what she felt was her calling as a messenger for God's creation.

When Sally was in seminary, she met with like-minded students to discuss how they could deepen the connection between ecology and faith. As Sally recalls, "Of course, saving Creation is central to most faiths, but until the middle to late 1990s few if any religious leaders so much as mentioned it."[10] Their discussions led to the formation of the Regeneration Project in 1994, a nonprofit whose mission was to deepen the connection between ecology and faith. But at first, they had a hard time raising funds, and they weren't quite sure what actions they might take.

It wasn't until a meeting of the Episcopal Church Ecological Network in Colorado in 1996 that the Regeneration Project ministry became clear. On a hike with an Episcopalian layperson from Massachusetts, Steve MacAusland, Sally learned about how certain states had deregulated the sale of electricity. In other words, instead of being stuck with one regulated monopoly utility from whom you had to buy electricity, customers could opt for different electricity providers that offered renewable energy. While it was more expensive to buy renewable energy at the time, if they coupled it with energy efficiency, churches could save money on their electricity costs while doing right for the environment.

For three years Sally, in California, and Steve, in Massachusetts, went church by church asking clergy to implement energy efficiency measures and switch to clean energy providers. They called their campaign Episcopal Power and Light. At that time, many religious leaders didn't want to touch the issue with a ten-foot pole. Some told Sally that the environment was a political issue and didn't belong in the church. She remembers feeling that environmental concerns were way too liberal for many congregations. They accused her of liking trees more than people.

"I was not popular, but I carried on," quips Sally.[11]

She remembers the reactions she got after preaching about climate change from the pulpit for the first time. She was told by the church leadership that she was driving people away. But it didn't stop her. She built up the courage, tried different approaches, honed her message, and continued talking about it. She realized that instead of blaming people or accusing them, she had to go at it slowly. She learned to teach the basics about our collective impact on the environment and did so repeatedly, until preaching about the subject became a real joy and she began receiving warm receptions.

She also learned how important the impacts of climate change on disadvantaged communities were to the faithful. Over time, events like Hurricane Katrina were able to demonstrate the reality of the situation. "The severity of Katrina and the profound evidence that poor and vulnerable communities are targets for disaster has brought climate change to the faith community as a matter of social justice, something religion has historically addressed."[12]

In three years' time, she had personally convinced over sixty Episcopal churches in California to switch to 100 percent wind power, and each church encouraged their congregants do the same. Fast forward to the year 2000. The California energy crisis hit. The experiment in deregulating California's energy providers had gone awry. Enron, and others who had gamed the system, went bankrupt, leading to a series of blackouts that left millions of customers without electricity. It also meant that the green energy providers that had come to California to offer renewable energy options could no longer do so. The clean energy providers had to leave the state, and Sally had to rethink her strategy. Without the choice to buy clean energy from a utility, Sally decided to focus on helping congregations reduce emissions with continued energy efficiency measures and conservation.

She also knew it was time to broaden the effort beyond the four hundred fifty Episcopal churches in California. To really make a dent in climate change, this needed to be an interfaith effort. At the same time, the demand for her work began to expand beyond the Episcopal Church. Denominations of every faith were seeking her guidance.

So she called a meeting in Oakland to gather the partners she had connected with over the years. Partners from the California Council of Churches, the Southern California Ecumenical Council, the Northern California Inter-Religious Conference, and the Coalition on the Environment and Jewish Life got together and decided to link arms. From that day on, they called themselves Interfaith Power and Light.[13]

While this work started in California, Sally knew that there was an increasing demand for it nationwide. Knowing that they couldn't be in all places at once, they created a network of IPL chapters. People of faith from around the country came to them wanting to engage locally. IPL started fundraising nationally and distributing that money to the local chapters. As news of the organization's work spread, there were soon IPL chapters popping up everywhere. Georgia, Minnesota, New York, New Mexico, and Connecticut were some of the early adopters. They started creating partnerships with the Environmental Protection Agency, including the Energy Star for Congregations program.

Today, IPL has over twenty-five thousand congregations in forty states that are installing solar panels, implementing energy efficiency measures, screening climate change films, lobbying their elected officials, and communicating the moral imperative of solving climate change from a theological perspective. Sally's message has been so widely adopted because it's simple and universal: "We will need to become conscious of our behavior," she says, "follow our faith's call to be stewards of Creation and believe in the power of the Spirit to move us toward a healthy planet. The goal is simple, the means are going to involve us all."[14]

GREENING THE BLACK CHURCH

One pastor in particular, my friend the Reverend Dr. Ambrose Carroll, realized that he could play a meaningful role in expanding the reach of Sally's important message. Carroll is an African American Baptist pastor from West Oakland. As a theologian, he was always looking for the "big tent, big umbrella issue of our day," something akin to the civil rights concerns of the 1960s. In the 1990s, while doing organizing work in the Bay Area, he met an up-and-coming civil rights organizer, our friend from chapter 5, Van Jones. Some years later, in 2007, Ambrose read Jones's book *The Green Collar Economy* and knew he had found it. Environmental justice was the issue he had been looking for. "Van articulated it well," Ambrose told me and RE-volv's Solar Ambassadors as the guest speaker on our webinar.[15] (RE-volv is a nonprofit I started in 2011 to empower people around the country to bring clean energy to their communities. You'll learn more about it in chapter 8.)

After reading the book, Ambrose knew he needed to get involved. He became a fellow with Jones's organization Green for All for three years, learning more about sustainability and the ways that it could benefit communities of color. During this time, he saw another way to bring the sustainability movement to people of color. While groups like Interfaith Power and Light were bringing the green movement to faith communities, and Green for All was bringing the green movement to the black community, there was still a gap. Someone needed to bring the green movement to the intersection of the two—the Black Church. It was critical to Ambrose that the church be included in preaching the green message, because "The reality is that the church has always been for us the major agent for community change."[16]

As Ambrose points out, there hasn't been a successful social movement to date that didn't have the Black Church at least at the table:

> At the end of the Civil War, it was Rev. Hiram Rhodes Revels of the AME Church, who served as an Army Chaplain and then fought for the rights of freedmen after the war, serving as a U.S. Senator from the State of Mississippi. It was the Black Church that held the center of leadership during the end of Reconstruction and fought against the perils of lynching by organizing the NAACP. It was again the church through the leadership of Martin King and many others that led during and after the Civil Rights Movement by founding such organizations as the Southern [Christian] Leadership Conference and C.O.R.E. The Black Church has always been a major agent of community change and still today has much untapped potential when it comes to ecological justice.[17]

In partnership with Green for All, Ambrose started a campaign, which grew into its own nonprofit, called Green the Church, which he continues to lead. The mission was bold: to bring green theology, sustainability practices, and political organizing tactics to African American churches around the country.

Ambrose explained to us that in the 1970s and 1980s, when the concept of environmental justice emerged, African American communities weren't thinking about it. They were focused instead on the traditional justice issues that plagued their communities: police brutality, gun violence, poverty, and the prison industrial system. But during recent years, leading organizations such as the NAACP and the Urban League developed environmental wings to address health issues like asthma and cancer that disproportionately affect communities of color.[18]

And there's a reason the environment seems foreign to many members of the black community, Ambrose told us. "There's a historical narrative that has separated us from the land," he explained.

African Americans were brought from one continent to another still very connected to the earth. While enslaved, they worked in agriculture, with their hands in the soil. After the Civil War, freed slaves asked for forty acres and a mule. "Our intent was to still be with the ground and with the earth," Ambrose said. But the Great Migration severed that connection. After World War I and through World War II, freed slaves left the farms in the South to find work in urban centers elsewhere in the country, where they worked, for example, as laborers building ships in shipyards and as assemblers in automobile factories. Soon, working in agriculture, with your hands in the soil, was looked down upon among urban African Americans. Farming also had negative connotations that traced back to slavery.

According to Ambrose, this separation is still felt among black people today. Laughing, he said, "We're not trying to go camping. Grandma ran away from Mississippi and Alabama. We're not trying to pick cotton no more. Not trying to be outside." He described an experience being outside with a friend's dog: "I was scared of the dog. In my DNA are generations of my people running away from dogs. Dogs take on a different meaning viscerally," he said. "For some, being in nature, being around trees, being in God's country is easy. But when your people were lynched from trees, that's not your thing anymore. We have to get past some trauma to get back to the created order—back to the animals, the trees, farms—and feel good about it."

Part of his mission with Green the Church is to break down these barriers. "I think there is a lot that we have to do just in terms of building ceremony that brings African-American people in particular, and all people, back in relationship with the land," said Ambrose.[19] He told us that the key to successfully doing so was "education, education, education." "As an African American pastor," he said, "when I get up and talk about sustainability, I get blank looks." For a community that's been through so much trauma, there's a lot of mistrust of outsiders. Ambrose's fear is that when good opportunities present themselves, like sustainability initiatives, there's still so much distrust that African American communities may miss out.

In order to create trust, you need to find a common language, Ambrose explained. Green the Church is using the language of the Bible, the language that rings true in his community, to heal collective traumas, and reconnect people with the earth. "If our people are truly going to move forward from a dirty energy economy to a clean energy economy, if our people are going to strive in a green collar economy, one that creates jobs that are pathways out of poverty, then the church must set the example, and the church must lead the way."[20]

Ambrose also wants to make sure that the green message is not just focused on economic opportunity but also on the moral imperative the reality of the situation demands. "If we continue to live how we are living," he said, "as if we have infinite resources, and as if we can use this world up and find another one to live on, we are in for a rude awakening." I couldn't agree more. As the report *Limits to Growth* articulated so well back in 1972, a planet with finite resources cannot sustain an economic model based on the myth of infinite growth.[21]

Green the Church is still in its early development, but the potential to activate the "sleeping giant that is the Black Church," as he describes it, is tremendous. Ambrose now serves on the board of directors of Interfaith Power and Light and hosts a Green the Church conference every year, bringing together pastors from African American churches around the country to share best practices on how to preach the green word.

WISCONSIN GREEN MUSLIMS

Ambrose Carroll is not the only one who saw an opportunity to tie green theology to a community not typically included in the environmental movement. Huda Alkaff is the founder of an organization called Wisconsin Green Muslims. According to their mission statement, "Wisconsin Green Muslims (formerly known as the Islamic Environmental Group of Wisconsin), a grassroots environmental justice group formed in 2005, intends to educate the Muslim community and the general public about the Islamic environmental justice teachings, to apply these teachings in daily life and to form coalitions with others working toward a just, healthy, peaceful and sustainable future."[22]

I first heard about Huda at Solar Power International (SPI), the biggest solar conference in the country, in 2016 in Las Vegas. I was there to speak at a Department of Energy event. While I was there, I got to catch up with friends from Convergence Energy and Renew Wisconsin. Convergence is a solar installer in Wisconsin that RE-volv has been working with for years, and Renew Wisconsin is a leading nonprofit that's been driving renewable energy policy in Wisconsin for decades. My friends told me I had to connect with Huda.

Many of the solar projects RE-volv finances are for faith institutions. For example, we've worked with Interfaith Power and Light and Green the Church to put solar on the buildings used by their member congregations in a number of states. Now, I was learning about Wisconsin Green Muslims for the first time. Huda, as my Wisconsin friends told me, was going to mosques around Wisconsin and presenting to them about the benefits of sustainability and, in particular, going solar.

On my next trip to Wisconsin for one of RE-volv's solar ribbon cuttings, I made a point to meet Huda, who took me to her mosque, where we met with the executive director and talked about putting solar up on the building. That day I also learned more about Huda's journey. As far back as she could remember, she wanted to be an environmentalist. She studied chemistry and biology and later earned degrees in ecology and sustainable development.

In 2005, only a few short years after the September 11 attacks, when Islamophobia was high in Wisconsin, Huda started to think about using sustainability to bring people together. Then she came across the unifying principles of solar energy. In an interview with the *Forum on Religion and Ecology at Yale*, Huda shared statistics that support the idea that renewables bring people together: "In our work, we are tapping into the unifying power of solar energy and water. We found through several polls in 2016 and 2017 that solar energy has high approval ratings among people from diverse political, social, geographical, and educational backgrounds—nationally and in Wisconsin."[23]

In 2017 she conducted a study to determine how sustainability could improve the image of Islam among Wisconsites, sending online surveys to over five hundred people. The results indicated that her hunch had been correct: educating the public about Islam's support of solar energy helped shift people's opinions about the religion. When non-Muslim Wisconsites recognized that Islam, like other religions, shares the values of sustainability and caring for creation, it removed some of their preconceived notions about Islam and created positive associations that weren't previously there.[24]

Huda works tirelessly on the Wisconsin Green Muslim programs, which are many. One of their more popular programs is Greening Ramadan, which encourages Muslims to incorporate sustainability practices into their holy month. The initiative, which started in Wisconsin, has now been adopted by Islamic groups around the country. They also have a Faithful Rainwater Harvesting program, teaching congregations to collect rainwater. And of course, closest to my heart is their Faith and Solar Wisconsin program. To date, Huda has given over eighty presentations to faith communities across the state, reaching over 2,500 people.[25]

Huda is also one of the founding members of the Wisconsin Interfaith Power and Light chapter and a founding member and leader of the Interfaith Earth Network. She serves on the national Interfaith Power and Light Campaigns Committee, the national Greening Ramadan Task Force, and the Milwaukee Environmental Consortium Board of Directors.[26] Safe to say, in case it wasn't clear already, she's a climate courage powerhouse.

KATHARINE HAYHOE AND EVANGELICALS FOR CLIMATE SOLUTIONS

I first learned about Katharine Hayhoe during an episode of the climate change miniseries called *Years of Living Dangerously* back in 2015. The program was made by James Cameron and features Hollywood celebrities investigating climate change. In this episode Don Cheadle visits Lubbock, Texas, to meet with Katharine, a climate scientist who is also an evangelical Christian.[27] Sound paradoxical? Perhaps, but only if we only believe the typical narrative that Evangelicals are politically conservative people who don't believe in climate change. But Katharine represents a growing number of Christians who recognize their role in caring for creation. Not only that, as a leading climate scientist and expert on the subject, she's using her platform to evangelize the faithful about the realities of climate change. What makes her work even more impactful is that she's using her experience to teach climate advocates how to better communicate the message of climate change.

I first met Katharine in 2016 in Leesburg, Virginia, at an Audubon conference. During her talk she pointed out that our belief in climate change has more to do with what tribe we belong to, what persona we identify with, than with anything else. For example, one of the slides she showed that day pointed out that of all the religious groups that accepted the reality of man-made climate change, the very top of the list was Hispanic Catholics. This made sense, as Pope Francis had recently released his encyclical on climate change. However, on that same slide it showed that white Catholics were the last group on the list. Same faith, same good book, same pope, but as far apart on climate change as you could be. What gives?

I had the chance to chat with Katharine briefly after this talk and twice more after talks she gave in the Bay Area. Even though she is talking about as serious a matter as climate change, Katharine remains upbeat, positive, funny, and optimistic. And even though she is one of the world's top climate scientists, she speaks in everyday language that everyone can understand. It's her relatability that makes her such an effective communicator. Her message is clear and startling: people's beliefs on climate change are more associated with their political ideology than anything else.

In America, our political ideology can be a strong aspect of our identity. It signifies our values, our tribe, where we belong. In her TED Talk, Katharine points out "if people have built their identity on rejecting a certain set of facts, then arguing over those facts is a personal attack."[28] Trying to tell them the latest science only makes them dig their heels in more. But if you talk about common interests, like love for the outdoors, with someone who doesn't outright dismiss climate change but whose views on climate change are doubtful, cautious, or disengaged, there's room to build common ground. And if you point to how changes to the climate are affecting the places that they hike through, go birdwatching in, or canoe in with their grandchildren, then reducing the threat of climate change to protect these areas will align more closely with who they are and what they care about. As she put it in her TED Talk,

> If you don't know what the values are that someone has, have a conversation, get to know them, figure out what makes them tick. And then once we have, all we have to do is connect the dots between the values they already have and why they would care about a changing climate. I truly believe, after thousands of conversations that I've had over the past decade and more, that just about every single person in the world already has the values they need to care about a changing climate. They just haven't connected the dots. And that's what we can do through our conversation with them.[29]

What Katharine is describing here is exactly what the Heath brothers and Simon Sinek suggested we do in chapter 2. Start with the *why*. Try to make your message resonate with the Elephant, the emotional brain. Connect the dots about climate change and things they already care about, so it's something they can *feel* is important to them too.

Katharine, who hails from Canada, began her journey by studying astrophysics at the University of Toronto and graduating with a bachelor of science degree. During her studies she took a class on climate science, taught by Danny Harvey, who had studied under the renowned climate scientist Steve Schneider. Her knowledge of astrophysics put her in the perfect place to understand the changing dynamics of our climate. And as she learned more about the magnitude of the crisis, her faith compelled her to devote her time to addressing the challenge. This decision led her to the University of Illinois at Urbana-Champaign to get an MS and PhD in atmospheric science. Now the director of the Climate Center at Texas Tech University, Katharine has published over 125 peer-reviewed works and was a lead author of the second and third US *National Climate Assessment*, while she continues to teach and do research.[30]

Her husband, Andrew Farley, is an evangelical pastor of Lubbock Bible Church and best-selling author of eight books on faith. They even penned a book together—*A Climate for Change: Global Warming Facts for Faith-Based Decisions*. They decided to write this book to help address the questions that came from "friends who wanted to know the 'truth' about climate change from a fellow Christian, someone they could trust."[31]

As an evangelical Christian, Katharine can connect with other people of faith about the importance of climate change. Because of her faith, she's able to speak to the faithful with honest sincerity in a language they understand, and she's been able to open hearts and minds in the process. Other Christians can look to her and know that it's OK to talk about climate change. And even other scientists can look to her and think it's OK to talk about faith. Part of what makes Katharine so successful in reaching people is that her message is simple. One of the biggest problems, she points out, is that even though the majority of Americans know that climate change is happening, two thirds never talk about it, and three quarters say they never hear about it in the media.[32]

Hayhoe translates what scientists know intimately into dinner table conversation, or Sunday morning church-coffee-hour talk. One of the obstacles to people understanding climate science more broadly is that what the scientists know doesn't often reach the public. Scientists sharing their findings with politicians, who pass it through the media, creating a giant game of telephone that only results in confusion and distortion of the facts. Katharine, guided by her faith, is making her message about the realities of climate change accessible and available to everyone. She is brimming with climate courage. What I've learned from her is that the most important thing we can do about climate change, which I'll keep coming back to throughout the book, is talk to other people about it, listen to their perspectives, and see if we can broaden the conversation.

The good news for the climate is that she's not the only Evangelical who's concerned about climate change. Among others, there are two organizations for which she serves as a scientific advisor that help champion the cause: Young Evangelicals for Climate Action and its parent organization, the Evangelical Environmental Network. In keeping with Katharine's advice to approach climate change from common interests and a shared set of values, the Evangelical Environmental Network has tied climate change to the pro-life stance that many Evangelicals share. Of course, the issue of a woman's right to choose is one of the most contentious issues facing our country. Reading this language might trigger your Elephant, the emotional brain, to run for the hills. But what makes it such a compelling example is that this group is relating climate change to an issue their members feel passionately about—an issue at the core of their identity.

The campaign is called the Pro Life Clean Energy Campaign, and its stated goal is to organize half a million pro-life Christians to participate in taking action for clean energy.[33] This group has taken the core concerns of climate change and the need for clean energy and tailored its language to speak directly to evangelical pro-lifers with conservative leanings who support free enterprise and limited government intervention in their lives. The petition on their website reads:

> As a pro-life Christian, I believe pollution harms the unborn, causing damage that lasts a lifetime. Dirty water and air have serious consequences for the health of our children and other vulnerable populations, like the elderly.
>
> So, I ask my Governor and other elected officials to support a plan for clean electricity that will: free our children from pollution by relying entirely on clean electricity from renewable resources like wind and solar by 2030; defend our freedom to create our own electricity from sunshine, without fees championed by monopolistic utilities; free our communities from regulations that prevent us from joining together to create our own electricity; and free businesses from such regulations so that they, too, can create and [sell] clean electricity.[34]

Note the language in this statement. It's framing the argument for clean energy with key conservative values. This might as well be coming out of Rush Limbaugh's lips:

- "As a pro-life Christian"
- "pollution harms the unborn"
- "defend our freedom"
- "without fees championed by monopolistic utilities"
- "free our communities from regulations"
- "free businesses from such regulations"

Note what's missing:

- Catastrophic climate change
- Melting ice caps
- Polar bears
- Al Gore

You get the picture. In order to reach across the aisle and have meaningful conversations about climate change, which Katharine Hayhoe and I both believe is the key to solving climate change, we have to learn how to frame things using the language that someone can relate to and identify with, not what we relate to and identify with. Remember, if you want to reach someone, you first have to connect with their Elephant, their emotional brain.

SHIFTING THE CONVERSATION

Religious institutions in the US still have a long way to go to help truly shift the conversation about climate change in their congregations and communities. There needs to be a deep reflection on the core values of their traditions, with a good faith effort to put aside the politics. It's certainly easier said than done. But thankfully, leaders such as Pope Francis, Sally Bingham, Ambrose Carrol, Huda Alkaff, and Katharine Hayhoe are showing us the way.

At the end of the day, climate change is a moral question. The actions that we take now to curb climate change, or not to, will have grave implications for the lives of others today and for millennia to come. For the countless species with whom we share this planet who are going extinct at alarming rates. For the future generations that will be born into a world whose climate was altered due to our current activities. And, front and center, for the members of vulnerable and disadvantaged communities in this country and around the world who are least able to deal with the effects of climate change, who have done the least to cause climate change, and who are most affected by climate change.

The great spiritual traditions of the world disagree on many things, but there are some things they all have in common: A tradition of serving others, and a tradition of treating others the way we would want to be treated. Caring for our neighbor. Offering a helping hand to those in need. Creating a positive impact through selflessness. If ever there has been a challenge that required this type of moral courage, this level of dedication to serving other people and all life on Earth, this degree of commitment to leaving the world better than we found it, climate change is it.

And as history has shown us, the moral conviction of faith communities has always played a significant role in the public discourse when moral questions arise. I take comfort knowing that these leaders are pushing their faith communities to step forward in the name of our common home.

CHAPTER 7

ENERGY INDEPENDENCE

We cannot drill and burn our way out of our present economic and energy problems. We can, however, invent and invest our way out.

—VAN JONES[1]

ONE THING THAT all Americans can get behind is energy independence. Whether it's to protect ourselves from other nations controlling our energy sources, to reduce involvement in foreign conflicts, or to keep our energy dollars here at home and benefitting our communities, Americans across the political spectrum favor increasing our homegrown sources of energy. The military, leading businesses, and parts of the country that are most vulnerable to extreme weather and energy supply shortages know that the best way to reduce their exposure to risk is to ensure that they have authority over their own energy supply. Thankfully, renewable energy makes that easier.

THE DEPARTMENT OF DEFENSE

No one is more aware of the importance of energy independence than the US Department of Defense (DoD). The US military is the single largest consumer of energy in the world. Stop. Let that sink in. It accounts for 80 percent of the energy consumed by the US government.[2] Its massive energy footprint means that its risk exposure to even the slightest price fluctuation or lack of supply could be catastrophic, especially when lives are on the line.

When it comes to energy security, there's no greater concern for the military than on the battlefield. American soldiers are deployed to all corners of the globe, to remote locations, and often in hostile enemy territory. That means access to energy is one of the greatest security threats that soldiers face. Hauling energy sources to the front lines is an extremely risky endeavor. In the early years of our wars in Iraq and Afghanistan a single Marine combat brigade burned through five hundred thousand gallons of diesel fuel every day.[3] The vast majority of supply trucks in a convoy were needed to carry fuel.[4] A 2009 report estimated that one out of every twenty-four fuel supply convoys resulted in a casualty because of the risks associated with transporting fuel.[5] Between 2003 and 2007 over three thousand American soldiers or contractors were killed in Iraq and Afghanistan while doing so.[6]

This tragic loss of life prompted the Marines to change their strategy. They assessed their options and found a winning solution: mobile solar photovoltaics. Instead of transporting energy sources via convoys along predictable routes, making them easy targets, Marines started carrying portable solar panels with them to forward operating bases. And unlike loud diesel generators—which can be heard for miles and allow adversaries to easily pinpoint their location—solar panels don't make a sound.[7] Now soldiers could air condition their insulated tents, run their computers, communication devices, and lights, all with a few solar panels and batteries, saving the military tons of money, but most importantly saving lives.[8]

The Department of Defense's focus on energy security by necessity extends beyond the battlefield to its supporting operations around the world, which include three hundred thousand buildings located in five hundred fixed military installations (bases, posts, and centers under control of the military) around the world.[9] Each branch of the armed services has set goals for quickly switching to renewable energy. And, thankfully, they're all well on their way to meeting their renewable energy targets. The Army and Air Force share the goal of 25 percent renewable by 2025; while the Navy is shooting for 50 percent by 2020. From 2007 to 2015 alone, the military reduced its oil consumption by 20 percent.[10]

The Department of Defense knows it's time to get the US—and frankly the whole world—off fossil fuels and is using its buying power to help make that happen.

Specifically, it has three main strategic concerns: First, how much of the DoD's time, money, and energy is spent on keeping the shipping of oil secure. Because of the importance that Middle Eastern oil plays in the world economy, the US military runs a massive operation to make sure that those shipping lanes remain safe—guarding the Persian Gulf with aircraft carriers has cost taxpayers an estimated $8 trillion since 1976.[11]

The next two strategic reasons to support a shift to renewable energy are born from the DoD's concerns about climate change. Second on their list of concerns are the deleterious impacts that climate change is already having on our nation's defenses. In a 2019 report to the secretary of defense, the Pentagon lays out the challenge succinctly: "The effects of a changing climate are a national security issue with potential impacts to Department of Defense . . . missions, operational plans, and installations."[12] The report points out that *over two-thirds* of all of our mission critical military installations around the world are threatened by climate change.[13] The Pentagon investigated seventy-nine mission critical bases across the Army, Navy, and Air Force and found that fifty-three installations currently experience recurrent flooding, forty-three face drought, thirty-six are exposed to wildfires, six are undergoing desertification, and one is dealing with thawing permafrost.[14] These are not future risks but are security threats currently impacting our armed services caused by climate change.

Lastly, the Pentagon has made clear, since as far back as the 1990s, that in addition to limiting our own defenses, climate change is a top security threat because of the impact it will have on people throughout the world. Climate change is a threat multiplier for all unstable regions of the world, due to the havoc that changes in the climate have on people's livelihoods and on the natural resources that ensure their survival. In a 2015 report to Congress, the Department of Defense found that

> climate change is an urgent and growing threat to our national security, contributing to increased natural disasters, refugee flows, and conflicts over basic resources such as food and water. These impacts are already occurring, and the scope, scale, and intensity of these impacts are projected to increase over time.
>
> . . . A changing climate increases the risk of instability and conflict overseas, and has implications for DoD on operations, personnel, installations, and the stability, development, and human security of other nations.[15]

It certainly makes you wonder, doesn't it? How do elected officials and media pundits still claim to be uncertain about the realities of climate change when the Pentagon has been putting out report after report declaring it one of the biggest threats to our national security for the last three decades? And how can they claim to want to keep America secure yet ignore the DoD's warnings about climate change? The good news is that, despite whatever president is in office or party in power, the Pentagon knows that reducing its footprint and preparing for the effects of climate change are critically important for the safety of our country. Hopefully its clean energy efforts will remain largely unaffected by politics in Washington.

BUSINESSES TAKING ACTION

When it comes to mitigating the risks of climate change energy price fluctuations, there's another powerful group that's not waiting for Washington: the business world. Most leading businesses across nearly all sectors of the economy recognize that the world is going renewable, and they want to be the leaders benefiting from the transition, not the laggards left out. Amazon. Apple. Bank of America. BMW. Citi. Coca-Cola. Facebook. GM. Google. Goldman Sachs. IKEA. Johnson & Johnson. Kellogg's. Lego. Lyft. Microsoft. Nestle. Morgan Stanley. Nike. Salesforce. Sony. Starbucks. Unilever. Visa. Walmart. Wells Fargo. What do these companies, whose products range from cars and food to tech and financial services, have in common? They've all committed to being powered 100 percent by renewable energy.[16]

In fact, over 160 major companies have made the renewable energy pledge through an initiative called RE100. Five of the six highest-valued companies in the world—Apple, Google, Microsoft, Amazon, and Facebook—have all committed to 100 percent renewable energy. On the one hand, they're doing the right thing by the environment and are acting in line with their principles as responsible corporate actors. But the other motivation is that it's good for the bottom line. Tech companies have massive data centers around the world with countless servers that are holding together our digital universe. And, not surprisingly, they require a lot of electricity. In the United States, the cost of electricity goes up each year, often between 3 and 7 percent. For a big company that uses a lot of power, that's a huge liability. To reduce the risk of uncertain electricity rates over time, these companies are opting for price certainty by purchasing their own renewable energy systems or signing long-term fixed price contracts known as power purchase agreements (PPAs) with renewable energy power providers. The advantage with renewable energy is that you can lock in a low rate for decades, or if you buy the equipment outright, you don't have to pay for the fuel ever again.

Typically, the transition to renewables works something like this. First, a company reduces the amount of energy it uses by improving efficiencies across its operations, which saves energy and money. Then it may add solar panels to its warehouses and other facilities, if it has a big roof like a Walmart or IKEA. In addition to putting solar on the roof, companies such as Apple, Microsoft, and Google need so much power that renewable energy developers have to build massive new solar and wind farms big enough to meet their power needs.

The costs of clean energy technology have dropped over the years, and we're now at a point where going 100 percent renewable is the wisest business decision. And thankfully, the more these industry leaders lead by example by switching to renewable energy, the more it will become accepted business practice and influence other companies to follow suit. Microsoft has actually set a new bar for corporate sustainability with its 2020 announcement to go beyond carbon neutrality. Microsoft plans to be carbon negative on a yearly basis by 2030, and by 2050 they plan to "remove from the environment all the carbon the company has emitted either directly or by electrical consumption since it was founded in 1975."[17] Wow. When a company makes a commitment to doing something that's never been done before and sets in place an ambitious plan to do so because it's the right thing to do, that's leading by their values—that's climate courage.

In addition to these companies reducing their own massive carbon footprints by implementing energy efficiency measures, building new solar and wind farms, and exploring other innovative strategies to draw down emissions, like massive tree-planting efforts, they're also setting the bar for what other businesses should do and how they should operate. On the renewables side, not only do these efforts result in significantly lowering carbon emissions today, but those large capital investments in renewable energy create a boon for the growing industry, further turning the clean energy flywheel.

THE BRONX BOMBERS

Even my hometown team, the New York Yankees, is getting on board with sustainability. Over the last few years, sustainability has been a major focus for the Yankees, which makes me smile with pride. Their efforts include energy efficient LED lights at Yankee Stadium and a zero-waste policy that keeps 85 percent of the stadium's trash out of landfills. They, along with a few other sports organizations, are also devising creative ways to offset the carbon emissions they haven't yet been able to avoid producing.[18] To keep their new sustainability initiatives on track, the Yanks hired an environmental science advisor, which is said to be the first position of its kind in any professional sport.[19] They're hoping it starts a trend.

In April 2019 the Yankees joined the United Nations Sports for Climate Action Framework, "the aim of which is to bring greenhouse emissions in line with the Paris Climate Change Agreement and inspire others to take ambitious climate action."[20] Yankees managing general partner Hal Steinbrenner said,

> The New York Yankees are proud to support the United Nations Sports for Climate Action Framework. For many years the Yankees have been implementing the type of climate action now enshrined in the Sports for Climate Action principles, and with this pledge the Yankees commit to continue to work collaboratively with our sponsors, fans and other relevant stakeholders to implement the UN's climate action agenda in sports.[21]

What's great about organizations like Apple or Nike or the Yankees committing to sustainability is that their efforts go beyond the impact of reducing their footprint or even inspiring other organizations in their industry. The real game-changer is in the minds of the public. Think about all the Yankee fans who will be surprised to learn about their team's commitment to meeting the Paris Agreement goals, and how that might shape their own opinions on the UN climate accords. Or for all the athletes who wear Nike apparel and know that their gear was made using renewable energy. As opposed to sustainability being something that's foreign or fringe, when brands that we trust demonstrate that they want to do something about climate change, climate action becomes a familiar concept and, eventually, people see it as just the way things are done.

VIVA LAS VEGAS!

While the companies we've already discussed have certainly gone above and beyond to demonstrate their commitment to sustainability, this next company had to claw its way through regulatory barriers to achieve energy independence.

As I'm sure you're aware, Las Vegas gets plenty of sun. And the hotels and casinos there use lots of electricity to keep their establishments open around the clock. In 2016, MGM Grand, owner of thirteen casinos and resorts along the strip, decided to invest in achieving energy independence rather than be subject to the increasing costs of their electric utility, NV Energy.[22] They started by adorning the decadent Mandalay Bay Resort convention center with twenty-six thousand solar panels, making it the largest contiguous rooftop solar array in the country at the time.[23] This array produces enough solar energy to power one thousand homes every year.[24]

But the astonishing 8.3 megawatt system still only covered 25 percent of the power MGM needed.[25] To further achieve energy security, they had to build more solar offsite. This posed a problem to NV Energy, the regulated monopoly utility provider. In most parts of the country, you can't choose the company you buy your energy from. To keep the electric grid functioning smoothly, most states are regulated, meaning they only allow one major electric utility to provide energy for everyone in a certain area. The original idea behind the model was understandable: state governments didn't want different power companies building different grids connecting their own wires every which way and creating a veritable mess. Today, however, many states have successfully deregulated their electricity markets, allowing customers to choose among different electricity providers all through the same wires.

But in Las Vegas, MGM didn't have that option. And NV Energy didn't want to let them go, as MGM's thirteen properties accounted for 7 percent of NV Energy's electricity sales. The three-member Public Utility Commission that regulates NV Energy set the fee that MGM would have to pay to source its own energy at $126.5 million.[26] The regulators justified the exit fee with an argument known in the industry as the "utility death spiral."[27] If one customer leaves the grid to go solar, then the costs of maintaining the grid are shared by fewer customers, which increases their rates, making it more attractive for other customers to leave and go solar too, creating a spiral that could put the utility out of business.[28]

This term was coined by the leaders of the utility industry, who recognize the real threat that solar and storage represent to their business model. For over one hundred years regulated utilities have had the luxury of being monopolies in their communities and never having to compete. Now solar and increased battery storage capacity pose a legitimate existential threat to the utility business model. While some utilities across the country are more solar friendly, many, like NV Energy—or Georgia Power or Florida Power and Light, as we learned from Debbie Dooley in chapter 4—are fighting solar tooth and nail, which is understandable—they're scared. The truth is, collectively, we need to come up with new business models that allow utilities to remain financially viable and to continue to provide essential grid management services in the age of solar and batteries. How to do so is a question that is hotly debated.

After negotiations, MGM settled on paying $86.9 million to leave NV Energy to source cleaner electricity elsewhere. In a letter to the Nevada Public Utilities Commission, MGM executive vice president John McManus said, "It is our objective to reduce MGM's environmental impact by decreasing the use of energy and aggressively pursuing renewable energy sources. Our imperative is heightened by increasing customer demand for environmentally sustainable destinations."[29]

MGM then partnered with solar developer Invenergy to build a 100-megawatt solar farm on 640 acres about twenty-five miles north of Las Vegas, part of a plot of land set aside by the federal government for renewable energy production. MGM is now proudly able to cover over 90 percent of the electricity needs of their thirteen properties on the strip, which include the Mandalay Bay Resort & Casino, Circus Las Vegas, the Bellagio, Mirage, Aria, and MGM Grand resorts.[30] (If you've stayed at one of these places recently, did you realize that your vacation was powered by the sun?)

What's so inspiring to me about this story is MGM's leaders' persistence. They put up the largest contiguous rooftop solar installation in the country, but they didn't stop there. Even when they were threatened with a $126.5 million fee to source electricity on their own, they continued to pursue their energy-independence strategy and were successful. The most telling aspect of the story, though, is that the economics of solar are so good that, even considering an $86.9 million exit fee, it was *still* good for MGM's bottom line and even better for its environmental footprint.

MGM Resorts' sustainability chief, Cindy Ortega, makes the point that in addition to reducing the company's footprint, in addition to saving money on electricity costs for years to come, they also have an opportunity to lead by example. "We have the ability to educate a wide variety of stakeholders on how we can exponentially reduce environmental impacts," she said. "Las Vegas is the perfect place to do that because we have 40 million people come here every single year, and so what better place to start telling that story."[31]

Now, that's climate courage. And telling that story is exactly what we need to be doing to shift our collective narrative about climate change.

THE ALOHA STATE

So far, we've examined the need for energy independence from a national security perspective, from a cost saving perspective, and as a sustainability leadership practice. Now let's talk about energy independence from a *resiliency* perspective. As a chain of islands remotely located in the Pacific Ocean far from the mainland United States, Hawaii faces great challenges when it comes to energy. For years it has been importing mostly oil for its electricity generation, which as you can imagine, creates some of the highest electricity rates in the country.[32] In 2018, electricity in Hawaii cost two to three times as much as it did in the rest of the US.[33]

Costs aside, from an energy security perspective, this also poses a tremendous risk. What if, due to a foreign conflict, a weather-related event, or a market shortage, the flow of oil stops or increases in price dramatically overnight? Relying on regular boat shipments of imported oil is an inherently high-risk approach to powering an island over the long term. As costs continued to rise, Hawaii explored its options. The solar irradiation (the amount of sunshine) in Hawaii is bountiful, making it a great place for solar energy. And because electricity costs are so high, Hawaii was the first place in the country where the cost of solar was lower than the grid electricity, known as *grid parity*. This led to massive solar adoption by residential customers, giving Hawaii the claim to the highest per capita solar adoption for homes and businesses in the nation.[34]

However, in 2015, due to the high adoption of distributed solar among its residential and commercial customers, the Hawaiian Electric Company (HECO) utility had to discontinue its net metering policy.[35] Net metering is the mechanism that allows you to get credits on your electric bill for every excess kilowatt-hour that your solar system generates and sends back to the grid. In most parts of the country, because the cost of electricity is lower, it doesn't always make economic sense to own solar panels and battery storage yet (although the economics of batteries are getting better by the minute). It's better to stay connected to the grid and to sell your excess power back to the utility. But after net metering was removed in Hawaii, customers without batteries would have been giving their excess power to the utility for free. Given the high cost of grid electricity, paying for batteries there, to store their own power, makes economic sense, and more and more residential customers are adding batteries to their solar power systems as a result.[36]

Thankfully, the state is taking the lead on renewable energy development and is the first in the nation to commit to 100 percent renewable energy,[37] which I'll discuss more in chapter 8. What I love about this story is that it shows how communities are taking control of their own energy security.

PUERTO RICO

Despite the progress we're making toward a clean energy future, the CO_2 we've already emitted into the atmosphere over the last two hundred years is already changing the climate, and we're already feeling the effects. While the hope of this book is to mobilize swift action to avoid worsening the situation, it is true that we are already locked into a certain degree of planetary warming, and that means we must prepare for those effects. As we've seen in countless climate-related disasters, from Superstorm Sandy to the Camp Fires in California, our electrical grid is often vulnerable to damage that leaves thousands without power.

There is no better example of this than in Puerto Rico, where, in 2017, Hurricane Maria decimated the electrical grid. Efforts to rebuild the infrastructure took nearly a year, during which time 1.5 million people were without electricity.[38] We've known for some time that our centralized electrical grid is vulnerable to extreme weather events. And as extreme climate-related events become stronger and more frequent, we have to start creating more resilient, democratized, community-based methods of delivering electricity.

Let me give you an example of how this works. Television is unidirectional. A news program is created by the TV station and is broadcast for everyone to watch. The internet, on the other hand, is multidirectional. Platforms like YouTube allow anyone to create their own content and share it with others. The same shift is happening with electricity. The centralized monopoly utility model is unidirectional. A big power plant generates all the electricity and inefficiently transmits it over long power lines to communities and people's homes. But the decentralized electricity model, made possible through distributed clean energy technology like solar, batteries, and electric vehicles, allows energy to flow in a multidirectional way. Homeowners could be both consumers and producers of energy, creating a surplus of clean energy every day that they share with their neighbors and that they are compensated for. Instead of just buying electricity from the utility, you would buy and sell electricity with your neighbors.

Environmental thought leader Jeremy Rifkin, in his book *The Third Industrial Revolution*, describes it like this:

> The emerging Third Industrial Revolution . . . is organized around distributed renewable energies that are found everywhere and are, for the most part, free—sun, wind, hydro, geothermal heat, biomass, and ocean waves and tides. These dispersed energies will be collected at millions of local sites and then bundled and shared with others over intelligent power networks to achieve optimum energy levels and maintain a high-performing, sustainable economy. The distributed nature of renewable energies necessitates collaborative rather than hierarchical command and control mechanisms.[39]

Rather than sprawling, inefficient and highly vulnerable electricity grids that still use one-hundred-year-old technology, what we're now seeing develop are smarter, localized, microgrids. Community-scale microgrids that are self-contained, powered by local clean energy resources like sunshine and the wind and maintained by ample battery storage are the wave of the future—not only from a resilience perspective but also a cost perspective. In contrast to how antiquated the current grid is, "at the very least, the smart grid may finally allow utility companies to know when the power is out—without receiving a phone call," writes David Biello in *Scientific American*.[40]

Puerto Rico is showing exactly how microgrids can work at scale. Now that so much of the electrical infrastructure has been destroyed, Puerto Rico has a clean slate to develop a new way of distributing energy that is more effective and more resilient. In early 2019 the Puerto Rico state legislature followed Hawaii's footsteps and adopted a bill that sets the goal of achieving 100 percent renewable energy by 2050.[41] Puerto Rico, like Hawaii, historically imported expensive fossil fuels to generate electricity and had transmission infrastructure that was particularly vulnerable. Along with the 100 percent goal from the legislature came an integrated resource plan from Puerto Rico's utility, Puerto Rico Electric Power Authority (PREPA), that lays out its strategy for the next twenty years. In addition to adding 2,220 megawatts of solar and 1,080 megawatts of battery storage, it plans to phase out coal and expensive fossil fuel imports.[42] In order to create a more resilient grid, the utility plans to divvy up the island into eight self-sufficient zones dubbed "mini-grids," which would be complemented by smaller microgrids in harder to reach areas.[43] Each mini-grid would be designed to ensure that local resources could provide power during and after major storms.[44]

Puerto Rico's goals to pioneer a new clean energy–based minigrid system that would encompass the island, will be an important case study. As the aging electrical grid in other parts of the US needs to be replaced, we can look to Puerto Rico's example to develop the clean energy mini-grids and microgrids that will power our communities into the future.

LEAPFROGGING

Internationally, clean, resilient, accessible energy has tremendous applicability as well. Currently over one billion people on the planet do not have access to regular electricity.[45] For people who live far from urban centers, often it's too costly for utilities to build out the electrical grid to reach them. Long transmission lines are an expensive, inefficient, highly vulnerable way to distribute energy. The same has historically been true of other technology; for example, in the early 2000s, when the cost of mobile phones dropped significantly, people in many parts of the world that did not have landline telephones leapfrogged to the new mobile technology, and those countries never built—and never will have to build—landline infrastructure. Now the world has more mobile phones than people.[46]

Similarly, clean energy generation and storage technologies on local microgrids are already leapfrogging traditional transmission technology in many parts of the world. Now, instead of families spending their money on costly, dirty kerosene for indoor lighting, leading to avoidable respiratory illnesses and fires, solar panels on the roof make an improved quality of life possible and also more affordable. I'll leave you with this: as the costs of these technologies comes down, and those one billion people who currently don't have electricity are able to experience the benefits of electricity that we so often take for granted, think about the incredible ways in which people's lives will change—solar power will give people access to clean water, refrigeration for medicine, light to study by at night, and vast economic opportunities. That's the world that clean energy makes possible by creating an abundance of energy accessible to everyone.

What I love about these stories is that resilience through energy independence is finally within reach. We don't have to wait for the federal government to bring clean energy to all. These companies, states, territories, and communities here in the US and around the globe can deploy clean energy right now, on their own, and save money doing so. They just have to decide to do so and get organized. The finances pay for themselves. And the benefits of having power in the case of an emergency is invaluable. As former New York City mayor Mike Bloomberg writes in his book with Carl Pope, *Climate of Hope*,

> The single most important development in the fight against climate change hasn't been the Paris Agreement . . . or even the advancement of solar and battery technology. . . . The most important has been that mayors, CEOs, and investors increasingly look at climate change not as a political issue but as a financial and economic one—and they recognize that there are gains to be made, and losses to be averted, by factoring climate change into the way they manage their cities, businesses, and funds.[47]

The good news is you don't need to be a mayor, CEO, or investor to bring the benefits of clean energy, energy independence, and resiliency to your home, neighborhood, or community. Now we all can do that. And that's very good news indeed.

CHAPTER 8

COMMUNITIES AT THE FOREFRONT

Since our leaders are behaving like children, we will have to take the responsibility they should have taken long ago.

—GRETA THUNBERG[1]

IF ACTING ALONE TO REDUCE our environmental footprint feels like it's not having a big enough impact and pushing for federal legislation from corrupt politicians seems like a pipe dream, where can we take action that feels productive and that helps to move the needle right now? Where can we truly be agents of change?

In our communities. Linking arms with each other.

The good news about climate change is that we now have the technology and means to implement clean energy and climate solutions on our own, in our own communities without relying on the help of big business or the government.

This chapter highlights inspiring people and initiatives doing just that.

THE POWER OF COMMUNITY

When it comes to making change in our society, we tend to think that if we make our voices heard and push our elected officials to act, they'll enact laws on our behalf that will make businesses change their ways for the betterment of society. But as we know, thanks to the $3.4 billion businesses spend each year lobbying those same elected officials, it's usually the other way around.[2] Big businesses get to change the laws in ways that suit them and their bottom line, and oftentimes at the expense of the rest of us. If our only hope is that each policymaker will one day grow a conscience or that the political sphere will rid itself of corruption, then I'm afraid we're going to be waiting a very long time. Much longer than our planet can wait for us to switch to renewables and decarbonize the economy.

Thankfully, there is another way we can create massive change that is often overlooked: directly influencing the market. Often characterized as voting with your wallet, this concept applies not just to your own purchasing habits but to the combined purchasing power of communities sustained over time. This strategy can send strong economic signals to the market, which in turn wields broad political, cultural, and social influence.

It's a tried-and-true strategy we've seen time and again, from the Boston Tea Party, to Gandhi's Salt March, to Dr. King's Montgomery bus boycott, to Cesar Chavez's Delano grape strike. Like boycotting goods and services from companies selling goods or employing practices we don't support, voting with our wallets can be a way to create positive change, by sending a signal for more of what we do want. As Maxine Bedat and Michael Shank explain in an article for *Fast Company*,

> Billions of green appliances are now sold every year. In 2018, green construction will account for 3.3 million jobs in America. Hundreds of thousands of electric vehicles are now on America's roadways. Solar panel installations on American rooftops have now surpassed 1 million. Organic food sales are seeing double-digit growth each year. That's consumer choice in action. By supporting greener products and greener companies and denying our support to environmental and social abusers, we can radically shape the way our companies behave. Businesses buy what we demand.[3]

With enough financial pressure from consumers for sustainable business practices, eventually, I believe the majority of industries will push our lawmakers to enact smart climate legislation, and the legislators will make it happen. In the meantime, it's up to us to continue to drive clean energy adoption and other sustainability practices on our own, at the community level, and in doing so create demand for the kinds of products and services that will steer the economy onto the path of sustainability.

For many years the costs of clean energy were so high that the only hope against climate change was federal policy. This is no longer the case. While Washington climate politics remain gridlocked, innovative community initiatives are taking advantage of the falling costs of clean energy to drive climate solutions locally without any help from Uncle Sam and, through their actions, further drive market trends.

THE VIRAL EFFECT OF SOLAR

Psychologists have identified a common behavioral pattern that can help us in the fight against climate change. It's called the "mere exposure effect." Chip and Dan Heath explain it in their book *Switch*:

> The more you're exposed to something, the more you like it. For instance, when the Eiffel Tower was first erected, Parisians hated it. They thought it was a half-finished skeletal blight on their fair city, and they responded with a frenzy of protest. But as time went by, public opinion evolved from hatred to acceptance to adoration. The mere exposure principle assures us that a change effort that initially feels unwelcome and foreign will gradually be perceived more favorably as people grow accustomed to it.[4]

This aspect of social behavior has the potential to make the biggest impact on the climate compared to anything else we do. And we can see that it's already playing out when it comes to solar. Around the country solar energy is proving to be contagious. It has a viral effect. Time and again it's been proven that if your neighbor has solar, you are more likely to go solar, and this results in solar clusters around the country that create ripple effects outward. In the *New Yorker*, Carolyn Kormann describes it like this:

> A recent study by researchers at Yale and the University of Connecticut found that socioeconomic and demographic factors like income, party affiliation, and the unemployment rate had little influence on the spread of residential solar-power systems. . . . The main factor that seemed to drive whether a household installed such a system was whether a neighbor had recently done so.[5]

Studies have shown that the same is true for energy efficiency measures.[6] While intuitively we might believe that the biggest motivating factor to adopting energy efficiency practices would be financial savings, the actual driver is social pressure from peers.

Over the coming pages we're going to look at some amazing organizations, changemakers, and initiatives at the local level that are planting the seeds of change in our communities and, perhaps more importantly, in our minds. Keep in mind as you read each story that the impact of these efforts goes far beyond the number of panels installed or kilowatt hours of clean electricity generated. As each innovator creates something from nothing or breaks through perceived barriers, our collective understanding of what's possible shifts. And as these stories are told, they inspire more changemakers, and thus we can create a snowball effect for positive climate action.

SOULARDARITY

Jackson Koeppel is a native New Yorker who went to college in Ohio. As a student he learned about the impacts of climate change and after seeing extractive fossil fuel practices in Ohio and West Virginia, he turned his attention to solutions. As a fellow in a Green Economy Leadership Training program working in an area of Detroit known as Highland Park, he discovered a particularly troubling situation.[7] He took a leave of absence from school to see what could be done.

Highland Park is a unique place. From 1913 until the 1930s, this city of three square miles (which is technically within the city of Detroit) was the home to Henry Ford's Highland Park Plant, where workers assembled automobiles. It was a lavish town, fit for Ford, his family members, and his executives. But today it's one of the most economically depressed cities in the country. By 2011 the city had accrued a $4 million electric bill with the utility company Detroit Edison (DTE). Due to delinquent payments, DTE removed all the streetlights in the city of Highland Park, and ten thousand residents were left literally in the dark every night. In a city that already had terrible crime and unemployment, the situation became a public safety nightmare.[8]

Jackson met with community members to see what could be done. After doing some research, they came up with an idea: to install streetlights powered by solar energy. To this end, Jackson cofounded a nonprofit called Soulardarity and started raising money through crowdfunding campaigns to pay for the streetlights. At this point they've installed six solar streetlights, which the residents own as a cooperative. As members of the cooperative, they each pay a minimum into the streetlight maintenance fund administered by the cooperative. Since the lights are stand-alone units equipped with batteries, they work even in the event of a power outage. In addition to providing light at night, they also create a mesh network that provides wireless internet to the community.[9]

Soulardarity is now working with the city government on a proposal to install one thousand further solar streetlights. Since the city is cash-strapped, the project has been put on hold for the moment, but in the meantime the cooperative is progressing one streetlight at a time.[10] Soulardarity is also working on organizing a group solar buying program, as well as coordinating weatherization financing for community residents.[11]

Through a program called SustainUS, which pairs young entrepreneurs with mentors, I had the great privilege of mentoring Jackson in 2014 and learning more about the work he does through Soulardarity. What I love most about this model is that it exemplifies self-reliance and community empowerment. The community identified a need, came up with a solution, and made it happen. They didn't have to go through the utility or the city to solve their problem. They just used their own ingenuity and wherewithal. Now other communities can follow the footsteps of Highland Park to secure their energy future.

RE-VOLV—MY PERSONAL ENTREPRENEURIAL JOURNEY

Another example of a community-led clean energy effort is RE-volv, the nonprofit that I founded in 2011 and where I serve as the executive director. In the summer of 2009, I went with my friend Michelle Ben-David to see a climate change documentary called *The Age of Stupid.* After the movie, we couldn't stop talking about it. Finally she said to me, "We have to do something about this." You could say that we both were already doing something about it. We were in our mid-twenties, and we were both working at environmental organizations (she was at the Natural Resources Defense Council, and I was at the Center for Resource Solutions) that focus on solutions to the problem of climate change. But she didn't think that was enough. She insisted that we needed to do something more. It stirred something in me. This idea that two young people could decide to do something about climate change outside of an existing organization or effort—that we could start something and be agents of change—that was pretty empowering.

The film made it incredibly clear that now is the time to act. What left us scratching our heads was this question: *What should we do?* What *could* we do that would have a meaningful impact? A week or so later Michelle and I met to discuss this further, and we called my friend Sean Miller to join the conversation. Sean was another longtime environmental advocate who, at the time, was working as the education director at the Earth Day Network. We each threw out ideas: A citizen-led campaign to promote public transportation. An environmental podcast that would feature important topics of the day. What we landed on was an idea I had been brewing for some time.

In 2008, in Gainesville, Florida, the local utility implemented the gold standard of renewable energy policy: the feed-in tariff. A feed-in tariff is a law that allows someone to build a renewable energy system and guarantees them the ability to connect to the grid to sell their power to the utility at a fixed rate. This policy mechanism was originally designed to spur clean energy development in the United States in 1978 under an act of Congress called PURPA (the Public Utility Regulatory Policies Act), but it only realized its full potential when Germany implemented it in the '90s under the energy minister and clean energy legend Herman Scheer. I had read Scheer's book *The Solar Economy* in 2008 and was pretty convinced that deploying this strategy was the best way to move clean energy forward quickly. After witnessing the success Germany had spurring the massive adoption of clean energy in the '90s and early 2000s through a robust feed-in-tariff, many clean energy advocates in the US agreed that it was a path we needed to follow. Gainesville Utilities was one of the first local utilities to enact it.

When I heard about Gainesville, I had an idea: Could someone purchase land and put up a ton of solar panels, then sell that power to the utility through the feed-in tariff and use that money to pay for another solar project? And then repeat that process over and over again? In other words, create a revolving fund for solar energy?

There were several other factors that played into my thinking back in 2009. The first was that solar energy was finally able to be financed. Jigar Shah, founder of SunEdison, famously devised the power purchase agreement, which allows someone to buy solar power as they go. Until then, people had to pay up-front for the full cost of a solar energy system that would provide them with twenty-five years of power. That would be the equivalent of paying for the next twenty-five years of cell service when you buy a cell phone. No one would have a phone, if that were the case. The reason everyone has a phone is that we can pay as we go. The power purchase agreement and other similar ideas, such as the solar lease, were being popularized at the time by companies like SolarCity, Sunrun, and Sungevity. In 2012 an article appeared in the *New York Times* called "The Secret to Solar Power," featuring my mentor and friend Danny Kennedy. The article describes the rapid growth that Sungevity experienced with the pay-as-you-go solar finance model.

The second factor that inspired me was the internet phenomenon Kiva. Started by Matt Flannery, who is also a friend and mentor, Kiva allowed entrepreneurs in the developing world to crowdsource small loans from people all over the world. Individuals who cared about fighting poverty were able to lend $25 to an entrepreneur trying to get a small business off the ground. The site had massive success and, with over a million lenders on the site, has transferred over a billion dollars in loans.

At my job at the Center for Resource Solutions I also saw that droves of people were signing up for green pricing programs through which they would pay extra for green electricity from their utility or purchase Renewable Energy Credits (RECs) or carbon offsets for a fee. People were clearly willing to pay extra to support clean energy.

The last factor that contributed to my thought process was learning about a revolving fund in action. In the book *Chasing the Sun*, Neville Williams, founder of Solar Electric Light Fund (SELF) and Solar Electric Light Company (SELCO), described a model used to bring solar to the Dominican Republic. Pioneered by a nonprofit called Enersol started by Richard Hansen, Enersol lent a few hundred dollars to a family to put solar panels on their roof. As Enersol was paid back over time, that money was lent to another family.

Williams describes it like this: "Richard formed a US nonprofit, found a few private donors, got a $2,000 grant from USAID, and set up an office outside Puerto Plata on the north side of the island. The Martinez family bought the first system on sight with a down payment and a three-year loan from the revolving fund Richard set up. Solar finance was born. By 1990, Enersol's community solar program had financed over 1,000 customers."[12]

The concept of a revolving fund is not new, but as I thought about using it in the US for community-based solar projects, a light bulb went off. I shared my thinking with Michelle and Sean. What if people could donate money to help build solar projects through crowdfunding? What if money earned from one project could help fund other projects? In other words, we could create a people-powered revolving fund for solar energy to benefit communities. As we talked, we came to the conclusion that what we were thinking of was like Kiva for Solar. Right. So instead of starting a new organization, why not see if Kiva could do it? I went to talk to Matt Flannery about it.

(On a side note, the original concept for RE-volv was closer to Kiva's, in that people could lend their money to solar projects and take it out of the RE-volv platform as their money was repaid to them. But since the costs of the solar projects were paid back over twenty years, we realized that would be too long for the average online investor lending twenty-five dollars. What we learned from Kiva was that the majority of people never took their money out of the Kiva platform but kept reinvesting it. So, we decided to pivot our model. Instead of asking people to lend us money, we would ask for donations, which would fuel a revolving fund.)

I had gotten to know Matt through an organization called blueEnergy, where he served on the board, and I had worked installing solar panels and wind turbines in unelectrified villages in Nicaragua. When we got in touch, I told him that I thought that there was a huge opportunity to crowdfund solar energy projects here in the US and that Kiva was well positioned to do it. He essentially said that Kiva's focus was alleviating poverty internationally and that adding US solar projects to the mix would be too much of a departure from that goal. He told me to go for it myself. So I did.

I wrote up a two-page document that laid out the concept and shared it with Michelle and Sean. I knew I wanted to start RE-volv, but I was afraid to quit my job since I needed the income. Then something magical happened: amidst the economic downturn, in 2010, CRS laid off me and a few other people. I felt like it was a clear sign from life telling me what direction I needed to go. I decided to take advantage of the moment by putting my head down and focusing all my efforts on launching RE-volv.

I asked Sean and Michelle to join me at RE-volv. But Michelle had started a new job and was preparing to go to law school. She didn't have the bandwidth. Sean was having a huge impact on climate change already where he was working at Earth Day Network and asked to join the board instead. I enlisted a few other former colleagues, professors, and friends to join the board, and we were off to the races.

I filed our articles of incorporation on February 4, 2011. Our fiscal sponsor was blueEnergy, which let us share an office with them. Mathias Craig, the executive director of blueEnergy, mentored and coached me and gave me worlds of valuable information about how to get a nonprofit off the ground.

I started waiting tables to support myself. I worked on RE-volv during the day and worked dinner and weekend shifts at a restaurant called the Plant Café, which serves organic food and, fittingly, was solar powered.

I worked with lawyers to figure out how to draft a solar lease agreement, and I worked with web developers to build our website. While many specifics still needed to be worked out, the organization's purpose was clear: to empower people to take meaningful action on climate change in their communities. I was sick of seeing documentaries like *The Age of Stupid* and walking out of the theater depressed. Even worse is at the end of the movie, when you're shown a short list of small strictly individual things you can do to affect climate change, such as put air in your tires and replace your light bulbs with more efficient ones. Once you're aware of the devastating impacts of climate change and the deep resistance to addressing it from industry and government, being told to put air in your tires or change your light bulbs feels like being told to spit on a forest fire.

Conversely, it is equally disempowering being told that the only real opportunity to effect change comes every four years when you can vote for the person who is the most concerned about climate change. As I've discussed, while Democrats and Republicans alike have acknowledged the reality of climate change, and some politicians have tried very hard to stop it, the US government is a slow-moving institution, and fossil fuel interests hold a lot of the power to influence it. Our federal government has not shown that it's capable of leading the clean energy transition, even though governments in other countries have, including, for example, Germany and other parts of Europe, and now China and India. I came to the conclusion that the most practical way forward was for people in their communities to start working together on solutions that were affordable, available, and scalable, like solar.

As we developed the model, it became clear that RE-volv could do three things with one solution: empower, invest, and educate. The first pillar of RE-volv's work was to *empower* people to take action. As at Kiva, I thought, through crowdfunding, RE-volv would allow anyone to donate a few dollars and know that their action resulted in more solar panels being built. That's pretty empowering. We also wanted to empower volunteers who had more time available to get involved in the process. So we started our Solar Ambassador program to train volunteer college students and community members around the country to help nonprofits in their community go solar. The volunteers would run a crowdfunding campaign to pay for the upfront costs of the solar energy system, which involved making a crowdfunding video, hosting events, engaging their networks on social media, and getting their story covered in the news. This has proven to be an extremely empowering process for our Solar Ambassadors. Recently, however, our model has changed because we've partnered with tax equity investors who put up the initial investments for the solar projects. Our Solar Ambassadors no longer need to run crowdfunding campaigns. Now they can exclusively focus on outreach to nonprofits, community education, and policy advocacy.

Invest is the second pillar of RE-volv's mission. Solving the climate crisis primarily boils down to this: investing a lot of money in building clean energy and other climate solutions. While financing exists for many homeowners and businesses looking to go solar, nonprofits are often left out. The federal government offers a tax credit for going solar, but that doesn't help nonprofits that don't pay taxes. It's also hard for nonprofits to prove their creditworthiness. These hiccups often make it too much of a hassle for traditional solar financiers to offer financing to nonprofits. Within that space, it's even harder for small-scale nonprofits that don't need a very large solar array. We realized that this was a gap in the marketplace that we could fill through our revolving-fund finance mechanism, the Solar Seed Fund—a pay-it-forward model that grows over time. As each nonprofit solar project is financed, the monthly payments from that nonprofit will help at least two other nonprofits go solar, and those two will help four more go solar, and so on. In my opinion, this type of strategy, which employs compounding interest to finance more and more solar (and could be applied to other climate solutions), is exactly the type of scalable effort we need to be prioritizing.

The other reason we chose to focus on solarizing nonprofits is because of the visibility they have. This brings us to our third pillar:

educate. As I've noted, one of the best things solar has going for it in the fight against climate change is that it's contagious. If you see solar on your neighbor's house, you are more likely to go solar, and this viral effect has been helping raise solar adoption rates in communities around the country. We wanted to know how nonprofit organizations, which already play important roles in their communities, could help more people see the benefits of solar. Our answer has been events, storytelling, and community engagement.

For example, when our Coastal Carolina University Solar Ambassador college fellows led a campaign to solarize VFW Post 10804 in Little River, South Carolina, it had an impressive impact. The veterans are mostly conservative, and Little River is a conservative town in a conservative state. Not to mention it was the first VFW in the state and one of the first in the country to go solar. When the students and the VFW leadership hosted a ribbon-cutting celebration and turned on the solar system, community members came, elected officials came, even the local TV station sent a crew to capture the event. A five-minute segment aired on the local news about the VFW that had gone solar to save money, so they can better serve veterans in need and also do their part to protect the environment. These are the types of stories that can change hearts and minds. And in fact, that's exactly what happened. A year later, nearby VFW Post 10420, in Murrells Inlet, South Carolina, went solar with RE-volv, following the example of their fellow veterans in the town over, with the help of our dedicated student volunteers. And our friends at Vote Solar also told the VFW's story in an op-ed, during a clean energy policy battle in the state, to help showcase the benefits of clean energy and win support in the legislature.

A few years back, Rev. Ambrose Carroll of Green the Church, who we met in chapter 6, introduced me to Rev. Curtis Robinson of Faith Baptist Church in East Oakland, California. Faith Baptist is an incredible African American church that has been a pillar of the community for over forty years. In a neighborhood that experiences high amounts of poverty and crime, the church is a respite that provides programs and services that support and uplift, including their signature food giveaway program that provides more than one hundred tons of food a year to people in the neighborhood who suffer from food insecurity. At the time, RE-volv was still crowdfunding the upfront costs of each of our solar projects. In this case, the Leonardo DiCaprio Foundation had offered to match every donation to the campaign. Plus, when the campaign went live, Leonardo DiCaprio tweeted about it. Our campaigns usually last six weeks, but this one, not surprisingly, hit its fundraising target in six hours.

At the ribbon-cutting event to celebrate the solar installation at Faith Baptist, the church was packed. Pastors and parishioners of other churches came to celebrate the inauguration and blessing of the solar panels, including our friends Rev. Carroll from Green the Church and Rev. Bingham from Interfaith Power and Light. Near the front row sat Deacon David Green of True Fellowship Baptist Church, an African American Baptist church in North Richmond, California, that plays a similar role in its community, providing food to the hungry and hygiene products to the homeless, among other services to those in need. While listening to Pastor Robinson tell the story of how Faith Baptist Church went solar, how they were motivated by their faith and their conviction to care for God's Creation, and how it was going to cut their electric bills in half, allowing them to provide more food for the hungry each week, a woman sitting behind Deacon Green tapped him on the shoulder and said, "We need to do this at *our* church." Walking out of the event that day, David gave me his card, and a few months later we were celebrating a solar ribbon-cutting at True Fellowship Baptist Church.

When a nonprofit that serves the community goes solar, it saves money and, as a result, is able to do more with its limited budget. As you see the direct benefits of solar energy in your community, you're going to feel more positively about it. This is the type of real-world application that will allow people to get past the rhetoric they hear on the news and realize that clean energy is something that works and benefits them and their neighbors directly.

Increasingly, our work at RE-volv is moving beyond education to activation. Once we get community members excited about clean energy, we want to mobilize them to take action. That might mean going solar at home, advocating for stronger clean energy policy in their area, or just telling their friends and family about how clean energy is benefiting their community. For example, at the ribbon-cutting events for Faith Baptist Church and True Fellowship Baptist Church, we had our partners from Interfaith Power Light attend to let parishioners know about their climate and clean energy policy initiatives and how to get involved. And our partners from GRID Alternatives came to let people know how they could help parishioners go solar at home.

To date, RE-volv has provided solar financing to over thirty nonprofits in California, Colorado, Illinois, Maine, New Mexico, Ohio, Oregon, Pennsylvania, South Carolina, and Wisconsin, collectively saving them over $2 million on their electric bills—money that they can now use to provide services to their constituents. In the process, we've helped these organizations to avoid emitting over eighteen million pounds of carbon dioxide during the life of the solar systems, which is equivalent to planting nearly seven thousand acres of trees, and in the process we've demonstrated the benefits of clean energy to over thirty thousand people in these communities.

GRID ALTERNATIVES

While RE-volv is serving nonprofits that can't get solar financing, there are still many other groups that are left out of solar financing options. GRID Alternatives, which I discussed in chapter 5, is a group I've admired for a long time. For over fifteen years they have been working to bring solar to low-income families. Not only that, but in the process they create pathways to clean energy jobs for those who need them most.

GRID was started by Erica Mackie and Tim Sears in 2001. They had been working as engineers on solar and energy efficiency projects and realized that there was a need to bring solar technology as well as the job opportunities it creates to low-income families. The idea was simple: train volunteers to get on the roof and install solar. While this process may sound rather complicated, GRID has developed a robust systematic approach to training volunteers on the technical aspects of solar installation, including how to carry solar panels, how to use tools properly, how to walk on a roof, and how to use proper safety gear. Today they've trained over forty thousand volunteers. While many volunteers are people who care about climate change and clean energy and are looking to roll up their sleeves and have an impact, the program's main goal is to train people who are looking for work to become solar installers. As I've noted, the solar industry is creating jobs twelve times faster than the national average. GRID is making sure that those job opportunities are available to everyone.

Because a lot of GRID's equipment is donated and because volunteers are building the solar arrays, GRID's model is to provide solar systems for low-income families at no cost. In 2004 they built their first system for a family in northern California. They continued to install solar for low-income families when, in 2008, the State of California started a program to do the same thing, called the Single-Family Affordable Solar Housing program or SASH. To ensure that low-income families were able to benefit from going solar, the state put millions of dollars into a program to pay for these solar systems for low-income families, and since GRID had been doing this work, they selected GRID to be the administrator of the program. At this point GRID went from doing a handful of projects a year to thousands. To date, GRID has installed over thirteen thousand systems, mostly in California but also in Colorado, the Mid-Atlantic, and Tribal Nations around the country.[13]

GRID is now the largest nonprofit solar installer in the country and has expanded internationally to bring solar to places like Nicaragua, Mexico, and Nepal.[14] If someone told you in 2004 that they wanted to give away solar systems to low-income families while training unskilled labor to install the systems, would you have dismissed it as a harebrained idea? I'm sure most people would have. And yet, Erica and Tim's climate courage, as well as the courage of all the founding members of the organization and all the families who trusted them to get on their roofs and install solar, and the government and industry partners that supported their work, have turned this into one of the most innovative and influential solar programs in the country, hands-down. Could you be the next Erica or Tim?

SOLAR EQUITY

As Majora Carter, longtime environmental justice strategist and founder of Sustainable South Bronx, says, "Environmental justice, for those of you who may not be familiar with the term, goes something like this: no community should be saddled with more environmental burdens and less environmental benefits than any other."[15] Yet we are far from that reality. Before I tell you about the climate courage of other amazing organizations, let's take a moment to discuss something that is often overlooked in the solar industry: equity. The industry is predominately made of white workers and disproportionately serves white communities. According to a recent report published by Tufts University and the University of California–Berkeley, African American communities are less likely to have solar than white communities by a factor of two thirds, and that's accounting for income disparity.[16] Similarly, the industry has yet to create an inclusive workforce. According to a survey from the Solar Foundation, the solar workforce is 73 percent white and only 26 percent women.[17]

This is exactly the type of eco-apartheid that Van Jones warned us about in *The Green Collar Economy*. Oil and gas refineries, drilling and fracking sites, chemical plants, toxic waste dumps, and other industrial sources of pollution are predominately located in communities of color, which have historically experienced the worst environmental impacts of modern society. Today, as Deborah Sunter, assistant professor of mechanical engineering at Tufts and lead author of the study, points out, "Unlike the fossil fuel industry, where energy injustice was attributed to exposure to negative consequences like pollution, with rooftop PV [photovoltaic] the injustice is more that certain communities are missing out on these economic benefits."[18]

We have the opportunity, with clean energy, to chart a new course. Because solar energy works almost everywhere, and its costs continue to decline, it has the potential to be ubiquitous. Its benefits—reduced cost of electricity, job creation, reduced local air pollution, resiliency during a blackout—should reach all communities. But that's not happening. There needs to be a concerted effort by the solar industry and the broader climate movement to include those who have been excluded. Thankfully, Sunter's study points toward a solution.

According to the report, "PV installations often result in a feedback loop: When a few residents in a community get solar, known as 'seed' customers, it compels others to join. Communities without those first-mover customers show delayed solar adoption."[19] As I've noted, solar adoption is viral. In fact, the study shows "when seeding does occur in communities of color, deployment 'significantly increases' compared to other racial or ethnic groups."[20]

The solar industry and the clean energy movement have to prioritize bringing these types of "seed" projects to communities of color, low-income and vulnerable communities and demonstrate the benefits and accessibility of solar. For example, RE-volv has partnered with Green the Church to bring solar to African American churches in Richmond and East Oakland, California, and has also brought solar to nonprofits that serve at-risk youth in communities of color in North Philadelphia, Pennsylvania, and East Dayton, Ohio. These projects are planting seeds for solar in the community. When GRID Alternatives puts solar on the homes of hundreds of low-income families in a community, that's planting many seeds. And every time Soulardarity installs a new solar streetlight that community members walk by every day, that's planting a seed for solar too.

Of course, these examples are just scratching the surface. The solar industry and policymakers need to do far more to ensure that all communities benefit from solar. Thankfully, there is a strong environmental justice movement that has been working on these issues for decades. Clean energy access is just one of the many issues they need to contend with. For example, the NAACP is getting involved in solar. They recently created the Solar Equity Partnership and are working with Vote Solar, GRID Alternatives, Sunrun, RE-volv, and others to ensure that beneficial solar policies and projects reach communities of color.[21] Other groups like Green for All, Asian Pacific Environmental Network (APEN), Communities for a Better Environment (CBE), the Climate Justice Alliance, Solar Richmond, and many more have been powerhouses for decades, fighting for their communities against dirty energy and making sure that they aren't left out of the clean energy future.

We need to make sure that history doesn't repeat itself and that the injustices of the last energy economy are eradicated in the new clean energy economy.

SOLAR UNITED NEIGHBORS

In college, I lived in a neighborhood called Mount Pleasant in Washington, DC, in a house where long conversations about how to solve the climate crisis and other social skills often went into the wee hours of the morning. Unbeknownst to me at the time, a few blocks away, a solar revolution was underway.

In 2007 Anya Schoolman's twelve-year-old son, Walter, came home one day with his friend Diego and told his parents that they wanted to go solar. They had seen the documentary *An Inconvenient Truth* and felt inspired to do something about climate change. As Anya looked into it, she found a lot of roadblocks. Keep in mind, this was in 2007, when the cost of solar was ten times what it is today. While her adult mind told her that there were too many obstacles, the climate courageous youth pushed her to think creatively, and together they came up with a model: the neighborhood "solar co-op."[22]

The idea was to get a number of neighbors to go solar at the same time with the same installer so they would all get a group discount. So, Walter and Diego walked around Mount Pleasant knocking on doors. Inspired by the climate courage of these two lads, forty-five families signed up for the Mount Pleasant Solar Co-op and installed solar on their homes. And it didn't stop there.

When the members of the Mount Pleasant Solar Co-op realized what they were capable of when they put their collective muscles to work, they decided that installing solar on their homes wasn't enough. Learning firsthand about the excessive amount of red tape involved in the process, they decided they wanted to make it easier for other DC homeowners to install solar too. They got involved in policy advocacy to reduce barriers to home solar installation, and they were extremely successful.

As people around the country learned about the innovative model and reached out for guidance on how to replicate it, Anya took what she had learned starting the Mount Pleasant Solar Co-op, packaged it as a toolkit, and shared it widely. To date, her organization, Solar United Neighbors, has organized solar co-ops in over a dozen states, has helped over 4,600 homeowners go solar, and has galvanized over one hundred thousand people to become solar energy advocates—and counting.[23]

The recurring theme in all these stories is that climate change is a winnable battle because so many of the solutions are simple, can be organized at the community level, and are easily replicable. Plus, we've got popular sentiment and economics on our side. All we need to do is give people an easy way to get involved, and the word will spread. What started as two twelve-year-old boys mustering enough climate courage to knock on doors in their neighborhood to talk to people about solar resulted in twenty-seven megawatts of solar deployed, laws changed in ten states, and nearly one thousand new solar jobs created. All because two twelve-year-old boys saw a film, were inspired, and had the guts to do something about it. Starting to feel courageous yet?

The organizations I've discussed so far in this chapter are building real clean energy projects that benefit people at the community level, that can garner public support, which is what's ultimately needed to make industries change direction and policymakers shift priorities.

THE INCREDIBLE HULK AND THE SOLUTIONS PROJECT

The actor who plays the Incredible Hulk in the Avenger movies, Mark Ruffalo, isn't just a green superhero on the big screen. He's also played a giant role in helping popularize the concept of 100 percent renewable energy. In 2011 fracking companies came to the area where Ruffalo lives with his family in New York's Catskill Mountains to explore fracking the Marcellus Shale, one of the richest natural gas fields in the world. As he learned about the damage that fracking natural gas was causing in small rural communities, he wanted to make sure that his family and community were protected. He became an outspoken leader in the fight against fracking, and the state of New York eventually banned fracking in 2014.[24]

In an interview with Earthjustice, Ruffalo explained:

> I was moved to step into the fight against hydraulic fracturing when I went to Dimock, PA and saw how their wells had been destroyed. I saw how crass and arrogant the companies who destroyed them acted toward their victims—refusing to take responsibility for the wrongs they had done. I saw that the local and state and federal government agencies that have been put in place to keep these kinds of things from happening were either apathetic or corrupt. I felt it was the right thing to stand up and say "No."[25]

He connected with clean energy business leaders, and they arrived at a question that many Americans have stumbled upon in conversations from board rooms to dinner tables: Can we really power everything with renewable energy? To explore this question, Ruffalo and his colleagues reached out to Stanford University professor of civil and environmental engineering Mark Jacobson, who I introduced in chapter 2, to see if he could help determine if New York State could be 100 percent powered by wind, water, and sun. What did they find out? You bet it can.[26]

Jacobson then expanded his research and found that the entire country could be powered by 100 percent renewable energy. He published a paper with his findings aptly titled "100% Clean and Renewable Wind, Water, and Sunlight (WWS) All-Sector Energy Roadmaps for the 50 United States."[27] Since the paper made a bold claim, and no one had ever done the research to back it up before, there was a lot of heated debate about it. But the evidence was there, and as costs have come down since Jacobson's initial research in 2015, it becomes more apparent every day that 100 percent renewable energy is in our grasp. In fact, he's since created roadmaps for 139 countries to be powered by 100 percent clean energy by 2050, which if implemented, would keep our planet below 1.5 degrees

Celsius of warming, the stated target of the Intergovernmental Panel on Climate Change.[28]

As a result of their advocacy, Ruffalo and this team of actors, scientists, and business leaders decided to start a nonprofit called the Solutions Project to help let the world know that the 100 percent clean energy powered society is possible. And they've been wildly successful in doing so. The organization is led by my friend Sarah Shanley Hope, who has done an incredible job growing it into the powerhouse it is today, successfully planting the 100 percent concept in the American imagination. This story caught the attention of another green giant: the Sierra Club, the nation's oldest and largest grassroots environmental organization.

READY FOR 100

The Sierra Club has been at the forefront of every major environmental issue over its 120-year history. Most recently, their Beyond Coal campaign, funded to the tune of $50 million by former New York City mayor Mike Bloomberg, single-handedly led to the closure of over half of America's coal-fired power plants (300 out of 580) through lawsuits that pointed out they were too costly to keep operating.[29] How's that for impact?

In 2016 they knew that it was time to throw their weight behind the 100 percent clean energy goal being showcased by Mark Jacobson's research. So they launched the Ready for 100 campaign. But knowing the challenges of working at the federal level to get our nation to go renewable, they have opted for a more distributed approach: they campaign community by community, county by county, city by city, and state by state to convince elected officials to commit to 100 percent clean energy before 2050. With thousands of volunteers across the US, the Sierra Club is one of the few grassroots organizations with a large enough reach to make a campaign like this work. The objective is twofold: to talk to people in communities around the country to find out how to craft local 100 percent clean energy plans that work for them and to get elected officials to commit to these plans; and to normalize the concept of 100 percent renewable energy in the minds of Americans.[30] Thankfully, as we've seen time and again, this simple strategy, easy to deploy, is working, and the message is spreading swiftly.

While not too long ago 100 percent renewable energy was considered out of reach, there are now commitments by over 160 cities, ten counties, eight states, and the District of Columbia.[31] And it's not just the usual suspects. These are commitments by places like Abita Springs, Louisiana; Cleveland, Ohio; Gainesville, Florida; Hanover, New Hampshire; and Houston, Texas. Hawaii was the first state to commit to 100 percent renewable. The second was California, which, if it was a country, would be the fifth-largest economy in the world. For California to commit to 100 percent clean energy means that renewable energy technology is no longer in the realm of fringe technology or supplementary energy sources but can be used as the sole provider of power to an economic juggernaut. And of course, as I discussed in chapter 4, since Governor Schwarzenegger implemented the country's first statewide cap on carbon in California, the state's economy has grown faster than almost any other state. Not wanting to miss out on the action, other states soon followed suit, making the 100 percent commitment. Now the list includes Maine, Nevada, New Mexico, New York, Puerto Rico, Virginia, Washington State, and Washington, DC.

Altogether, the commitments of these states and cities account for over 110 million people. That's one in three Americans who now live in a community committed to 100 percent clean energy by 2050.[32] And as I've noted previously, now 85 percent of registered voters want their communities powered by 100 percent clean energy. Let that sink in for a moment. A nimble coalition of volunteers around the country were given a simple training. They searched inside for their climate courage, found it, and collectively unleashed a wave of clean energy commitments across the country. I get so inspired just thinking about it. Don't ever let someone say we can't do anything about climate change. We can. We are doing something. This is what it looks like. It's easier than you think, and it's certainly worth the effort. As Margaret Mead taught us all, "Never doubt that a small group of thoughtful, committed citizens can change the world; indeed, it's the only thing that ever has."

UTILITIES GETTING ON BOARD

As the economics of clean energy improve and public demand for it increases, even some formerly recalcitrant utilities are getting on board. Xcel Energy provides power in Colorado, Minnesota, North Dakota, South Dakota, Wisconsin, Michigan, Texas, and New Mexico. All together they serve 3.6 million electric customers and 2 million natural gas customers, making them one of the largest energy companies in the country. They are now the first utility in the country to commit to providing one hundred percent carbon-free electricity by 2050.[33]

Xcel has been on the wrong side of the environment and the communities it serves many times. It's a for-profit company that's driven first and foremost by making a profit for its shareholders. The fact that Xcel has committed to going 100 percent carbon free is significant. First, it means that renewable energy sources, especially wind and solar in the Midwest and the West, are so plentiful and that the technology costs are dropping so quickly that Xcel realizes that renewable energy is the future of energy production. Second, Xcel, being a national industry leader, is also a trendsetter and their commitment could open the door for other utilities to follow suit. That takes a lot of climate courage, and I give credit to Xcel and to the advocates in those eight states who pushed the utility to make this commitment. But I suspect it's not just climate courage that is motivating them—it's also good economics. Even Arizona Public Service, which had long fought against renewables, has since followed Xcel's lead, and in January 2020 it announced a plan to retire coal by 2031 and reach 100 percent clean energy by 2050.[34]

Of course, the climate courageous are not always successful. David Crane was the CEO of NRG, which at the time was the largest electric utility in the country and whose primary energy sources were coal and natural gas. In 2009 Crane called renewables "a fact of life" and made a strategic shift for the company to become a leader in renewable energy generation.[35] While the long-term economics of Crane's vision were spot-on and already beginning to prove themselves, investors hadn't yet seen the writing on the wall. They couldn't tell that the dirty energy assets they owned were losing value in a world in which renewable energy will outcompete fossil energy. In 2015, during Crane's attempt to transform the company to primarily focus on renewable energy, investors lost confidence, share value dropped, and Crane was fired. The company still owns a sizable amount of renewable energy assets and will continue to invest in the space, just at a much slower pace than Crane had envisioned.[36] The fact that only three years later a company as large as Xcel has committed to going entirely carbon free, which even Crane had not proposed, speaks to how quickly the economics and politics of energy are changing in the United States.

To be clear, while most of the stories I'm highlighting are success stories, there are many examples of people exhibiting climate courage whose efforts ended up failing. Even as the solar industry was on a rapid growth curve, industry leaders like SunEdison and Sungevity went bankrupt, and SolarCity had to be purchased by Tesla to survive. That's the nature of a dynamic field. But whenever someone demonstrates climate courage and takes a risk the way David Crane did, they are creating change and paving the way for others to follow, often showcasing valuable lessons for the movement to learn from. Whether your climate courage efforts are wildly successful, fail miserably, or fall somewhere in between, if you are creative, thoughtful, and hardworking, your efforts will help solve the climate crisis. As Theodore Roosevelt put it, a quote Brené Brown named her book after, "The credit belongs to the man who is actually in the arena . . . who strives valiantly; who errs, who comes short again and again, because there is no effort without error and shortcoming . . . who at the best knows in the end triumph of high achievement, and who at worst, if he fails, at least fails while daring greatly."[37] Now is the time for all of us to dare greatly.

VOTE SOLAR AND THE IMPORTANCE OF LOCAL POLICY

I first met Adam Browning back in 2009. While I was working at the Center for Resource Solutions, I was helping to author a white paper for the Smart Electric Power Alliance (SEPA), an industry advocacy group, and I needed Adam's expert opinions on feed-in tariffs, net metering, and how various policy mechanisms could be implemented by utilities looking to expand their renewable energy generation. Adam is the solar policy expert who started the national organization Vote Solar, which works at the state level across the country fighting for nuts-and-bolts energy legislation that allows solar energy to prosper. To give you a sense of how important local solar policy is, allow me to share with you what I learned from another solar pioneer, Andrew Birch.

In 2017 Sungevity, one of the top three residential and commercial solar companies in the US at the time, went bankrupt. A few months later, its former CEO Andrew Birch, or Birchie, as he's lovingly known, did some soul searching and also some fact finding. He eventually penned an article that revealed a startling fact: solar energy was far more competitive in the European Union and in Australia than it was here in the US. Why? Not because of strong government mandates or more innovative solar companies. It was because of the crippling red tape and government bureaucracy in the US at the local level. The applications and paperwork required for siting, permitting, and interconnecting solar in the US double the costs we pay for solar compared to what our friends pay across the pond and down under. Not to mention that solar companies in the US have to navigate the complex policy landscapes of fifty states with dramatically varying regulatory frameworks for energy.[38]

This is the type of work that Vote Solar does. For example, in 2017 in Nevada, they worked with local solar owners to reinstate the solar bill of rights and make sure that homeowners would get fair credit for the solar energy they generated and sold to the grid. In 2019, in South Carolina, they worked with a coalition of groups to pass the Energy Freedom Act, to ensure a stronger renewable energy future by getting rid of a cap on the amount of energy that was allowed to come from solar.[39]

Clean energy policy can become tricky very quickly, as solar providers and advocates must navigate the interplay of utilities, state public utilities commissions, and energy goals and laws set at the state, regional, and national level. Thankfully, groups like Vote Solar analyze the technical details of various policy structures, mobilize constituents, and tirelessly advocate to push solar forward.

COMMUNITY CHOICE

One powerful policy tool that helps communities deploy more clean energy is community choice aggregation (CCA). Pioneered in California, CCAs have been implemented around the US and are halfway between a municipally owned utility and an investor-owned utility (IOU). If city or county residents vote to form a CCA, the city or county procures energy on behalf of its residents while the IOU maintains the transmission and distribution of the power. A CCA allows citizens to green their electricity mix through a democratic process no longer beholden to the utility. The CCA can purchase solar and wind energy from producers in the area on behalf of residents, but it doesn't have to manage the poles and wires as it would if the city completely municipalized.

What's really exciting is that the community can build its own clean energy resources over time—resources to be owned by the CCA. As the CCA begins to earn money, it can start to invest in building its own renewable capacity. It can put up solar panels on the rooftops of municipal buildings, over parking lots, and over reclaimed landfills. This "local build-out" strategy has a lot of advantages. First of all, it greatly reduces the future costs of energy for residents. If, for example, a community builds a solar farm, it will pay off the costs over a period of five to ten years and then begin to produce energy for free. It also creates local jobs. Lastly, this strategy makes the community significantly more resilient. As I discussed in chapter 7, Puerto Rico's localized energy resources with storage and microgrids are going to be increasingly advantageous for communities.

In the Bay Area, we've seen CCAs take off. Marin is home to Marin Clean Energy (MCE), the original CCA, which began in 2008. MCE had a lot of challenges, as pioneers often do. High premiums, pushback from community advocates for buying energy from solar farms far away, and, of course, local politics. But in the spirit of climate courage, it was also the first CCA in the country. That's no small undertaking, and the people who spearheaded it deserve a lot of credit.[40]

In San Francisco, the implementation of community choice has been a long drawn-out process. The plan that was first approved by the board of supervisors back in 2012 was held up in the public utility commission for more than five years. The delays were caused, in part, by the utility Pacific Gas & Electric's (PG&E) relentless attempts to thwart community choice efforts. They don't want to lose all their customers from San Francisco. But advocates kept pushing, and eventually Clean Power SF was rolled out.[41]

Over in the East Bay, a tireless group of advocates had for many years worked to launch their CCA, East Bay Community Energy. One of these advocates is my friend Al Weinrub. For years, the organization he runs, the Local Clean Energy Alliance (LCEA), has been one of the primary voices advocating for the creation of a CCA in the East Bay. Having learned lessons from Marin, San Francisco, and other cities that implemented CCAs, Al wanted to make sure that East Bay Community Energy was set up correctly from the beginning. This included advocating for having a strong local build-out component as a part of the initial design of the program and ensuring a high level of community participation in the decision-making process. Because without having the voices of the community at the table, how can anyone be sure that the CCA will deliver benefits to those it's intended to serve?

As with so many of the stories in this book, climate solutions rarely start out perfect. But after one community has made a go of it, makes mistakes, and learns lessons, their experience creates an opportunity for the next community to do it better, and as each one teaches one, we're collectively figuring out the path to sustainability.

COMMUNITY SOLAR AND THE SOLSTICE INITIATIVE

Here's another game changer: community solar. Also known as shared solar or solar gardens, this is essentially a subscription model for the energy from a local solar array built someplace in or around the community, other than a private home's roof. For example, a community solar project could be built over a parking lot, in a field, or on the roof of a community building. People from the community would sign up and pay money each month for a percentage of the energy produced by the solar system. Every month, through virtual net metering, the utility would credit the amount of energy produced through the remote solar system to the subscriber's electric bill.

The success of this model is critically important to ensure that the benefits of solar are accessible to everyone. Currently, if you rent rather than own your home or if your roof is shaded or isn't in good condition, you might not be able to have solar on-site. By some estimates this accounts for 25–50 percent of American houses. That means that people who care about solar energy won't be able to go solar at home. What's worse, of everyone who wants to lower their electric bills with solar, only a fraction of people will be able to do so if the solar panels have to be on their own roof. Community solar allows everyone, regardless of their living situation or roof condition, to be able to benefit from solar energy.

I first met Steph Speirs, a community solar pioneer and climate courage exemplar, in 2015 at a memorable place: the White House. As part of President Obama's efforts to support solar energy, the White House put together an initiative called the National Community Solar Partnership. The idea was to convene the groups doing innovative work on community solar around the country. While RE-volv's work isn't technically "community solar," because we work to bring solar to community-based nonprofits, we're often associated with community solar and are included in the conversation.

Steph and I were chosen to present together on a panel about our work. I was blown away by the work of her group, the Solstice Initiative. Based in Boston, Steph and her team have figured out how to offer the benefits of solar to community members who would typically be considered ineligible for such programs, due to low income or lack of credit history, through financial innovations and a digital platform that connects potential subscribers to local community solar developers.

Innovative models like Solstice Initiative are exactly the types of ideas we need to make solar accessible to everyone. But as I've noted, one of the challenges facing clean energy development in the United States is that the policy landscape tends to vary dramatically from state to state. While about a dozen states have enacted community solar legislation, the rules that regulate how community solar programs are set up and administered differ in each state, sometimes dramatically. This makes it challenging for companies in the community solar space to be able to truly scale. A few states—Minnesota, Colorado, and Massachusetts, for example—have great community solar legislation in place.[42] This is allowing for a handful of community solar companies to grow, but the potential market size of up to half the homeowners and all the renters in the country remains largely untapped.

CLEAN ENERGY ENTREPRENEURS

Of course, there's another community without whom very little progress would be possible. That's the community of clean energy entrepreneurs. As Steph Speirs could tell you, along with many of the entrepreneurs in this chapter, entrepreneurship is not for the faint of heart. Steph embodies climate courage so strongly and is so passionate about it that she's willing to give it everything she has to see her vision become reality. And she's one of the lucky ones whose business is growing and scaling.

The reality is that most entrepreneurs, whether in the clean energy business or any other industry, don't make it. Thirty percent of businesses don't make it past the two-year mark, and 50 percent don't make it past the five-year mark.[43] As Elon Musk has described it, "Starting a business is like eating glass and staring into the abyss."[44] In the clean energy world, it's arguably even harder. It's a nascent industry that's underinvested in, trying to disrupt the fossil fuel industry, the most profitable industry of all time, in a policy landscape that's fickle at best and outright trying to make the business model illegal at worst.

And yet the upside is like nothing we've ever seen before. Those companies that do succeed will be able to provide clean abundant energy to a world desperately in need of it. In fact, there has never been a business opportunity so compelling. This is why it's so important that clean energy entrepreneurs are given the support they deserve.

Danny Kennedy is a long-time Greenpeace activist turned clean energy entrepreneur. He started Sungevity with two cofounders and led it to become one of the largest solar companies in the country. Danny argues that for the industry to rapidly scale up, we're going to need thousands of entrepreneurs trying out new ideas and innovative technology, business models, scaling strategies, and marketing channels.[45] Obviously, only a handful will become the Googles and Amazons of the clean energy world. But we're going to need to try those thousands of business ideas in order to land on the game changers.

Emily Kirsch, who is now at the forefront of clean energy entrepreneurialism, saw an opportunity when Danny Kennedy mentored Mosaic cofounders Billy Parish and Dan Rosen to launch what is now one of the country's largest residential solar lenders. To help support a new wave of clean energy entrepreneurs, she partnered with Danny as a cofounder to start Powerhouse, a clean energy incubator and accelerator in Oakland, California.[46]

The classic tech start-up trope is two or three cofounders tinkering in someone's garage. An incubator is essentially a shared workspace that entrepreneurs can work out of together. They have a few desks, access to all the office amenities they might need, and a conference room for meetings. A start-up accelerator, also common in tech, is a place where scrappy founders trying to get an idea off the ground can be supported through coaching, mentorship, networking, and, potentially, some investment dollars. While such setups are common in the tech world, no one had created one for clean energy until Emily Kirsch and Danny Kennedy started Powerhouse.

Emily and Danny took an open space in the Sungevity office in Jack London Square in Oakland and initially called it the SFUN Cube, which stood for Solar for Universal Need. In addition to providing mentorship and connections to solar entrepreneurs, they also wanted to create an ecosystem of clean energy entrepreneurs who supported each other and who could learn from each other, partner with each other, and push each other. "We need to be courageous, irreverent, forward-thinking, and—most of all—positive," Danny wrote in his book *The Rooftop Revolution*. "To revolutionize our energy system for the good of our country and the world, we need leaders in the solar-energy movement to be firebrands and troublemakers, passionate activists, and savvy, scrappy entrepreneurs."[47] This incubator was designed to foster that spirit.

While Danny Kennedy went on to lead New Energy Nexus, a nonprofit that supports, connects, and invests in clean energy incubators and accelerators like Powerhouse all over the world, Emily Kirsch has taken the reins as CEO of the incubator and renamed it Powerhouse. Not only that but she's since turned Powerhouse into a fund itself that invests in select companies in the space. Like the tech ecosystem that was created in Silicon Valley, which birthed the major companies that dominate personal computing and the internet, Powerhouse has put Oakland on the map as the hub of clean energy entrepreneurialism in the United States.

I hope this chapter has made it clear that the technical solutions to climate change are already available to us. It's a matter of getting people on board in support of making the switch. And like any cultural shift, that doesn't come from the top down. It doesn't come from the government telling us what to do. It doesn't come from an industry pushing its products on us. It comes about when communities and groups of people support each other, find their collective climate courage, and take action. When they're successful, others will follow suit. If we make renewable energy a cultural trend and in turn a market trend, then businesses and lawmakers will start to take notice and try to adapt to the trends. Renewable energy and sustainability trends generated by the people can become the new norm.

Thankfully, community-led initiatives like the ones described in this chapter have gotten the ball rolling in the United States and are starting to build momentum. Now it's up to us, Americans who care about our country, who care about our children's future, who care about creating a more peaceful, equitable, and sustainable world, to join in the effort. Find what gives you climate courage, find a specific thing that inspires you, and grab a hold of it and give it all you've got. Because like so many of the folks described here, you might just help change the trajectory of our collective future.

While the deployment of clean energy technology and other sustainability practices is paramount in the fight against climate change, in the next chapter, I want to turn our attention inward, to something that may not seem as relevant at first glance as social activism or the infrastructure of renewable energy but perhaps is even more so—our values.

Environmental thinker Alan Durning writes, "The future of life on earth depends on whether the richest fifth of the world's people, having fully met their material needs, can turn to non-material sources of fulfillment; whether those who have defined the tangible goals of world development can now craft a new way of life at once simpler and more satisfying."[48] In other words, beyond the technical shifts we must pursue, what shifts in our minds, hearts, and culture, will get at the root of what's causing climate change?

CHAPTER 9

GRATITUDE, SIMPLICITY, AND SERVICE

We are here to awaken from the illusion of our separateness.

—THICH NHAT HANH[1]

ENVIRONMENTAL THOUGHT LEADER Gus Speth captured the societal moment we're in to a tee when he wrote:

> If today's growth and capitalism are delivering high levels of life satisfaction, genuine well-being, and true happiness to societies broadly, then there may be scant chance for real change. But if what we actually have is "spiritual hunger in an age of plenty," there is a large space for hope. A system that cannot deliver the well-being of people and nature is in deep trouble. It invites ideas and actions that are transformative.[2]

You may have heard the adage that the Chinese character for "crisis" is the combination of two symbols: danger and opportunity. Very often, our collective narrative about climate change is more focused on the danger it poses than the opportunity it presents. And in order to start shifting our collective actions toward sustainability, we're going to need to think of it the other way around, as the psychological research has shown. If people can see that they can improve their health, economic opportunity, community resilience, interpersonal relationships, and overall well-being by helping to create a sustainable future, then they will be inclined to do so. If instead, they think of climate solutions as an inconvenience or a bitter medicine that they must begrudgingly accept in order to preserve society, well, that's going to quickly discourage our collective Elephant.

As Speth points out, if people were generally happy and things were going relatively well for everyone, we might not have a good chance in the fight against climate change. People would cling to the current way of doing things. But many people in America feel our system has failed them, and many suffer from diseases of despair, the signs of which have been reported widely. In 2019 *income* inequality in America hit its highest level since just before the Great Depression[3] (with Jeff Bezos, Bill Gates and Warren Buffett collectively holding more wealth than 50 percent of the population combined[4]). As of January 2020, the opioid epidemic has claimed seven hundred thousand lives since 1999.[5] The suicide rate is higher than it has been for the last seventy years.[6] And the obesity epidemic now affects over 70 percent of Americans.[7]

According to a 2018 study by the World Economic Forum measuring happiness around the world, the United States, despite having the highest gross domestic product (GDP) in the world, also has the highest percentage of people taking antidepressants.[8] Despite our economic grandeur, the US shows up at number eighteen on the happiness list.[9] Economist Jeffrey Sachs, an editor of the report, reflected in an interview on the findings: "I think there really is a deep and very unsettling signal coming through that US society is in many ways under profound stress, even though the economy by traditional measures is doing fine. The trends are not good, and the comparative position of the US relative to other high-income countries is nothing short of alarming."[10]

It doesn't help that the average American spends over eleven hours per day listening to, watching, reading or generally interacting with media,[11] during which they see up to five thousand advertisements,[12] which studies have shown oftentimes lower our self-esteem.[13]

What's clear is that the media we consume or the products we buy can never replace the intangible parts of life that really give us happiness. Alan Durning eloquently describes the predicament represented in the World Economic Forum's findings:

> Psychological evidence shows that the relationship between consumption and personal happiness is weak. Worse, two primary sources of human fulfillment—social relations and leisure—appear to have withered or stagnated in the rush to riches. Thus, many of us in the consumer society have a sense that our world of plenty is somehow hollow—that, hoodwinked by a consumerist culture, we have been fruitlessly attempting to satisfy with material things what are essentially social, psychological, and spiritual needs.[14]

Psychologist Daniel Goleman, whom we met in chapter 2, echoes Durning's analysis, writing in his book *Focus* that "global economic data shows that once a country reaches a modest level of income—enough to meet basic needs—there is zero connection between happiness and wealth." He goes on to point out that "intangibles like warm connections with people we love and meaningful activities make people far happier than, say, shopping or work." Consumption, in other words, can never replace community. But in order to move away from our habits of overconsumption that drive climate change, we're going to have to take a long hard look at the values we live by as Americans and reflect on what needs to change. Thankfully, there's a field of psychology that can help. *Positive psychology*, a term coined by Abraham Maslow in 1954, is the study of human potential, how human beings thrive, and what contributes most to our well-being. We'll be hearing from positive psychologists Mihaly Csikszentmihalyi, Angela Duckworth, and Shawn Achor, whose research can guide us on the path.

In this chapter, based on the latest findings of positive psychology research, I suggest that, in order to address climate change, we need to cultivate different values—values that place a greater emphasis on community and less on consumption—and that living according to these values will have the benefits of reducing our impact on the planet and increasing our personal well-being. To do this I'll describe what I believe to be an effective three-step approach: (1) cultivate gratitude, (2) choose simplicity, and (3) focus on serving others. If we can learn to be more grateful for what we have, simplify our lives, and put more effort into serving others, I think we'll be well on our way to a happier, more sustainable world.

GRATITUDE

Our ability to be grateful can be one of the most important factors that determine our happiness. As the psychologist Mihaly Csikszentmihalyi reminds us, "Happiness . . . does not depend on outside events, but, rather, on how we interpret them."[15] This, he points out, brings us to the reasonable conclusion that "it seems that those who take the trouble to gain mastery over what happens in consciousness do live a happier life."[16] As the old saying goes, our happiness often boils down to whether we see the glass as half empty or half full.

Do I often look around and feel thankful for the number of good things going on in my life? Or do I look at what's missing, what's lost, or what hasn't yet been attained? Rather than constantly being in a state of want, in a state of lack, in a deprived, craving mentality because we choose to focus on all the things we don't yet have or haven't yet achieved, we can instead choose a different focus. We can focus on what we feel grateful for. Through mindfully guiding our thought patterns, we can cultivate the attitude that our lives are abundant, full, and that we have plenty of everything we need. We don't have to feel compelled to get a bigger home when the home we have is comfortable, cozy, and lovely in all its imperfection.

Wabi-sabi is a Japanese term that embodies the spirit of perfection in imperfection or taking pleasure in the imperfect. It stems from the idea that nothing is perfect, nor will it ever be. That piece of furniture with the slightly worn corners is perfect the way it is. It's the worn corners that give it that special something. A little character, perhaps. The key to seeing things this way is a gratitude mindset.

Save Money

By adopting a gratitude mindset, we can save a whole lot of money too. If I'm content with my home, car, phone, and other possessions, then I won't be tempted to have more or upgrade constantly. When we're not constantly shopping, spending, or upgrading the material things in our lives, we can actually save a lot. That means fewer credit card bills to pay, less stress and conflict about money in the family, less pressure to stay at the job you hate because the pay is good, and a greater ability to put some money away and feel more financially secure.

Less Stress

The biggest benefit of cultivating gratitude is that it's an antidote to stress. What are we most stressed about? Not having enough. Enough money. Enough time. Enough space. Enough praise. Enough love. Gratitude shifts our perspective. Rather than stressing about the annoying things about our jobs, we can be grateful that we are gainfully employed when so many are not. While we can certainly get overwhelmed sometimes by all the housework that needs to be done, we can choose to shift our focus to gratitude, knowing how lucky we are to have a place to call home.

Using Less

When I'm grateful for the clean water that comes through my faucet and I'm mindful of how precious a resource it is, I'm going to be more mindful not to waste it. I'll shut it off when I'm brushing my teeth or scrubbing a pot. When I'm grateful for all the energy and time that went into planting, growing, harvesting, transporting, and preparing this bowl of rice, I'll be more mindful not to let it go to waste. I'll save it for tomorrow or give it to someone else who might like it. In fact, one of the most potent ways we can fight climate change is by cutting down food waste. While so many around the world, and even here in the US, go to bed hungry, about 40 percent of the food produced in the US is wasted.[17] It's a shocking statistic, indicative of a culture that takes things for granted—in this case, food, one of our most fundamental needs.

Saying grace, like many religious practices, is less common today than it used to be. But what a beautiful way to start a meal—to reflect on how grateful you are for the food that you get to eat, for the people you might be eating a meal with, for your health, for all the people and all the processes that were involved in bringing this food to you. How might your meal taste different if this were to become a part of the ritual of eating? How might such a practice affect your relationship to food?

When we cultivate gratitude in our lives, we tend to switch our mentality from one of scarcity to one of abundance. It gives us a sense of confidence. A hop in our step. Coming from a place of abundance allows us to approach life more compassionately and less competitively, more collaboratively and less selfishly.

Generosity

If we know that overall we're doing fine, we're going to be more willing to help others. Even if we're not well off, even if money is tight, if we cultivate an attitude of abundance, we'll find that we're more often able to spare something to help someone else out. It's this type of generous, altruistic spirit that binds us together, that creates community.

Generosity also makes us happier. As the old saying goes: it's better to give than to receive. But now we actually have data that backs it up. Study after study shows that the giver is often happier than the receiver.[18] Perhaps we shouldn't be surprised then, by what happiness researcher and psychologist Shawn Achor and his team have found: a good deed has sort of a domino effect.[19] One person doing something kind for someone else will result in that person doing something kind for someone else, and so on and so forth. In the world today, in America today, I'm sure we all can agree that a little more kindness would go a long way.

How to Increase Gratitude

Despite a commonly held belief that brain damage and disorders are permanent and that human nature is unalterable, a number of recent scientific studies have demonstrated that humans have higher degrees of brain plasticity than previously imagined.[20] These results suggest that behavior patterns are less fixed than scientists used to think.

An example of brain malleability is known as the Tetris effect. If you're unfamiliar with Tetris, it's a video game in which the goal is to fit together different shaped blocks into a single row by moving and turning them. Psychologists ran studies in which they had people play Tetris for many hours in a row for three days in a row. What they found is that as the participants went back to the real world, their brains continued to see Tetris shapes everywhere they went. Their brains had become accustomed to looking for a pattern. Even though they were finished playing the game, the brain kept looking for it.[21]

We can use this same type of training to improve our mood and outlook, by cultivating gratitude. In fact, as Achor notes, "when researchers pick random volunteers and train them to be more grateful over a period of a few weeks, they become happier and more optimistic, feel more socially connected, enjoy better quality sleep, and even experience fewer headaches than control groups."[22] Psychologists have found a simple routine to implement that does just the trick: gratitude journaling. At some point during the day, write down three things you are grateful for. It could be anything: The breakfast you had. A compliment you received at work. A nice phone call with a friend. Getting through your to-do list. A beautiful sunset.

If you make paying attention to positive elements of your life a daily practice for a period of time, your brain learns the pattern. Since it knows that each day it will be asked to come up with three things to be grateful for, the Tetris effect kicks in, and all day long your brain will be looking for things to be grateful for. Thus, throughout your day, you will be feeling grateful for this interaction, grateful for that smile, grateful for the traffic clearing up, and so on. Studies have shown that gratitude journaling, done even for a short time, can have a lasting impact: "One study found that participants who wrote down three good things each day for a week were happier and less depressed at the one-month, three-month, and six-month follow ups."[23]

Once you begin this practice, you may find the number of things you're grateful for increasing dramatically. It also can help remind us of our interconnectedness. Who were the people who labored to construct the building or park you're in? Who were the architects who dreamed of it? We can be thankful for all these people and all of their efforts that allow us to be here enjoying this moment.

SIMPLICITY

The evidence shows us that what really gives us happiness are the very things we're not getting enough of: Social relations. Leisure. Community. Quality time with the people in our lives. Taking time for ourselves and enjoying the fruits of our labor. If social relations and leisure are important factors in our happiness, is modern society helping? As technology replaces social interaction, it increasingly isolates us. And ironically, with all our time-saving devices, it's actually speeding up the pace of our lives, leaving us with less leisure time to just be and enjoy life.

Simplicity at its core is about removing the extra activities, possessions, and responsibilities that don't bring us joy, so we can put more time and energy into the things that do. Perhaps there's no greater advice on simplicity than that given by the American author, philosopher, and father of the environmental movement, Henry David Thoreau. Thoreau, who spent two years, two months, and two days at Walden Pond observing nature, journaling, and reflecting on life, believed that even during his lifetime in 1840s Massachusetts, people needed to simplify their lives: "Simplicity, simplicity, simplicity!" he wrote. "I say, let your affairs be as two or three, and not a hundred or a thousand."[24] I can certainly relate. Unfortunately, I often feel as though the items on my to-do list number closer to a thousand than two or three. And frankly, I have a lot less on my plate than many people.

Simple living, however, does not mean we need to go live in the woods like Thoreau. Duane Elgin, author of the classic *Voluntary Simplicity*, argues that living simply is not about living in poverty or deprivation. It's about living an examined life in which one has determined what is truly important and how much is enough and then letting go of the rest.

Elgin's research led him to the conclusion that "the American public has experienced . . . limited rewards from the material riches of a consumer society and is looking for the experiential riches that can be found, for example, in satisfying relationships, living in harmony with nature, and being of service to the world."[25]

He goes on to say that the call to live simply is both pulling and pushing humanity at the same time. "On the one hand," Elgin says, "a life of creative simplicity frees energy for the soulful work of spiritual discovery and loving service—tasks that all of the world's wisdom traditions say we should give our highest priority." That's a strong pull. But there's also a push. "On the other hand," he points out, "a simpler way of life also responds to the urgent need for moderating our use of the world's nonrenewable resources and minimizing the damaging impact of environmental pollution." That's also true. "Working in concert, these pushes and pulls are creating an immensely powerful dynamic for transforming our ways of living, working relating, and thinking."

Declutter

Not surprisingly, in a time when clutter has become a serious strain on people's lives, Marie Kondo has become a famous author and TV star around the world by sharing her simple method for decluttering. Her techniques and sage advice help people get rid of the stuff that takes up space and mental energy. The KonMari Method, which she teaches in her books and on her TV show, asks us to tidy up the space we're in by removing what doesn't "spark joy."[26] Her message is simple: keep only the things that speak to the heart. By doing this, we can reset our lives and embark on a new lifestyle.[27] Given the levels of stress we all feel, her offer is quite appealing.

Decluttering our spaces by removing what no longer brings us joy frees us up to pursue activities and experiences that do bring us joy. "As a result," Kondo writes, "you can see quite clearly what you need in life and what you don't, and what you should and shouldn't do."[28] These dramatic changes in lifestyle and perspective, she argues, are life transforming.[29] I'm so glad to see that her message resonates with so many people, because this shift to simplicity is exactly what we need to reduce the overconsumption that contributes to climate change. What we find as we discard the material possessions that we've acquired over the years is that our joy and happiness don't reside in those things. It's a worthy reminder that material goods and possessions are rarely the things that bring us joy in our lives. If anything, they often get in the way.

Declutter Our Schedules

Americans work more hours than workers in any other Western country, take less vacation, retire later, have less parental leave, and set far fewer boundaries between work and personal time. We're more likely to eat lunch at our desks, to work late, to come in to the office on the weekend, and to process emails after dinner or first thing in the morning.[30]

Philosopher Alan Watts asks, What's the point of all our toil if we don't enjoy our lives? "If you say that getting the money is the most important thing," he says, "you'll spend your life completely wasting your time. You'll be doing things you don't like doing in order to go on living, that is to go on doing things you don't like doing, which is stupid."[31] Perhaps he's a bit heavy-handed, but I see his point. We have a choice in all of this, despite the societal pressure. More and more people are choosing to slow down. More companies are trying to create better work-life balance. More parents are trying to give their children more unstructured play time. Let's also embrace the value of simplicity for its own sake. As we slow down, take on fewer responsibilities and commitments, and declutter our schedule, we'll find we have more time for relaxing, more time for our loved ones, more time for ourselves, more time just to enjoy being alive.

And by the way, we'll create a lot less carbon than we do rushing about in our currently overscheduled lives.

Experiences

Money, it turns out, gives us a lot more happiness when it's spent on experiences, rather than on things.[32] So as we clean out our closets, getting rid of all the extra things we don't need, we should make a mental note: the next time we're tempted to make an impulse purchase, use that money instead on a fun excursion, a concert, or an evening out with a friend, which we'd be sure to enjoy more. Instead of working longer hours to afford more things, we'd be happier spending that time exercising, gardening, journaling, connecting with a loved one, reading, cooking a meal, or going for a walk.

Nature

Speaking of experiences that provide joy, few activities give us the mood boost and the health benefit of spending time in nature. Even short windows of fifteen minutes spent outside have vast impacts on our psychology and physiology.[33]

Time spent in the natural world also reminds us that we are part of the biosphere. The plants, animals, bugs, mountains, waterways, and woodlands nearby are an extension of our homes and part of our community. It's important for us to remember these natural areas surrounding us, to spend time in our local environment, and, hopefully, to cultivate a sense of stewardship in the process.

Spirituality, Creativity, and Self-Care

The ways in which people name and articulate their sense of the sacred come in more flavors than ice cream. And for some people the concept of a spiritual or mindfulness practice doesn't resonate. But for folks for whom it does, wouldn't it be restorative to spend more time in meditation, prayer, journaling, and reading inspiring words?

Meditation, which can help us cultivate mindfulness, is a technique taught in virtually all wisdom traditions. It is often as simple as sitting quietly and focusing on one's breath. But meditation can take all kinds of forms, including walking meditation, guided meditation, or any activity in which you can be fully present each moment.

The benefits of meditation and mindfulness practice are being studied by top universities and scientists at places like Harvard, Yale, and Johns Hopkins Universities, among others. What they're finding is that mindfulness meditation improves concentration and attention, reduces anxiety, helps with addiction, helps kids perform better in school, and even has been shown to increase cortical thickness in the parts of the brain associated with learning and to decrease brain cell volume in the area associated with fear and stress.[34]

For those whose spirit feels most alive when they're involved in creating, wouldn't it be nice to have more time for painting, writing, and making music? Many feel most vital and present when focused on the body—running, hiking, biking, doing yoga, or dancing. Whatever you do to stay grounded, simplifying our lives can give us more time and opportunity to be present and practice self-care.

Maker's Mindset

Mihaly Csikszentmihalyi is famous for his theory of "flow." He describes it as "the state in which people are so involved in an activity that nothing else seems to matter."[35] It's the feeling when "time flies" because you're so engaged in what you're doing.

Flow is a state of engagement in an activity that's both challenging, and gratifying. It "appears at the boundary between boredom and anxiety, when the challenges are just balanced with the person's capacity to act."[36]

For example, have you ever been cooking a meal, playing a game of chess, shooting a basketball, or painting and fallen into a deep trancelike state in which you forgot about anything but what you were doing? That's flow. According to Csikszentmihalyi, this is the state that our brains yearn to be in, and in fact, the more often we are in a state of flow, the happier we are.

Another type of activity in which people find flow, satisfaction, and presence is in making things. Humans are a crafty bunch, and we like to use our hands. Though we live in a society in which everything can easily be bought, we actually get far more enjoyment, far more value, and more appreciation from those things we create, build, or care for ourselves. How much tastier is your homemade salsa than the brand you buy in a store? How much better are your mom's chocolate chip cookies than those that come in a package? How much more do you like the hand-built table that your friend made you as a gift than the one you bought from a furniture store?

How much more satisfying is it to wear your favorite sweater that you patched up rather than replaced when you found a hole?

Not only does this maker mindset, this commitment to simplicity, this DIY mentality, save you money and reduce your environmental impact, but it also is more conducive to being in a state of flow. Learning new things and trying new activities like gardening, knitting, cooking with new recipes, repairing cars, and home improvements, can be extremely engaging and pleasurable. It also makes us more grateful for the things in our life when we realize the effort that goes into producing or maintaining them.

People

Science shows us, not surprisingly, that the area that gives us the most happiness, fulfillment, and meaning in our lives is our relationships. This is why rebuilding our sense of community is at the core of combatting the climate crisis.

How might we, as a culture, shift our collective values such that we can spend more time with our loved ones? What can we do as individuals to prioritize our family, our friends, our coworkers, our neighbors, our community members? How, when we're constantly bombarded with emails, pings, and updates that so often feel like they deserve our immediate attention, can we carve out more time to have meaningful conversations with the people we love the most? As the famous saying goes, no one on their deathbed ever says, "I wish I had spent more time at the office." They always wish they had spent more time with the people who meant the most to them. Simplifying our lives can help that happen.

SERVICE

The fast-paced, hyper-individualized materialistic world we've inherited was made by people. And as Frederick Douglass pointed out when analyzing the brutalities of slavery, "What man can make, man can unmake."[37] I was reminded of his words when I saw an inspiring poster at a bus stop recently that said "Tomorrow's World Is Yours to Build."[38]

Wow. That's a powerful message. And, frankly, very good news. I find it uplifting because it puts us in the driver's seat. *We* have the ability to reinvent the world.

The cultural narrative of our day is very much focused on individual development, individual power, individual fulfillment—in other words, on "me." My money, my stuff, my job, my image. Our individualistic culture is making us ever more anxious, ever more concerned with our appearance, and generally preoccupied with our life story—our personal narrative of the dramas of our day-to-day lives.

One way to break from this self-absorption is to direct our energy toward service. By service I mean serving other people, other living beings, for their primary benefit, and not our own. Not only do we end up better off by creating waves of positive impact that uplift our communities, but studies show that it makes us happier.[39]

Why does service make us feel happy? A lot of reasons.

Human Connection

First and foremost, serving others allows us to connect with people and strengthen our relationships to them. As social animals, we get a tremendous amount of joy and satisfaction from connecting with our friends, family members, coworkers, neighbors, and fellow community members.

As psychologist Angela Duckworth points out, from an evolutionary perspective we're hardwired to connect with each other. Human beings have evolved to seek meaning and purpose through serving others, she argues, because people who cooperate are most likely to survive compared to loners. Therefore, the desire to create meaningful connections with others in this way is one of our basic human needs.[40]

Purpose Driven

Nothing makes us feel more accomplished than helping others, lifting others up, and creating real impact in the lives of other people. I'm sure you've had plenty of these experiences.

Duckworth's "grit" research shows that the most successful people are not always the smartest or most talented but rather are those who, despite the obstacles, can persevere: those with the ability to persist and not give up easily. She writes:

> When I talk to grit paragons and they tell me that what they're pursuing has *purpose*, they mean something much deeper than mere intention. They're not just goal-oriented; the nature of their goals is special.
>
> . . . The long days and evenings of toil, the setbacks and disappointments and struggle, the sacrifice—all this is worth it because, ultimately, their efforts pay dividends to *other people.*
>
> . . . My guess is that, if you take a moment to reflect on the times in your life when you've really been at your best—when you've risen to the challenges before you, finding strength to do what might have seemed impossible—you'll realize that the goals you achieved were connected in some way, shape, or form to the *benefit of other people.*[41]

Serving others brings the best out of us and that feels great. And nothing feels better than giving your all, especially when it's for someone or something you really care about.

The Drum Major Instinct

In 1968 at Ebenezer Baptist Church in Atlanta, Georgia, the Reverend Dr. Martin Luther King Jr. gave a speech called "The Drum Major Instinct." His theory is that we all have an inborn desire to be great—to be first in our class and to receive praise. "We all have the drum major instinct," he said. "We all want to be important, to surpass others, to achieve distinction, to lead the parade. . . . This quest for recognition, this desire for attention, this desire for distinction is the basic impulse, the basic drive of human life—the drum major instinct."[42]

Dr. King points out, however, that the culture in which we live funnels that desire into a selfish, competitive pursuit of trying to outdo the Joneses.[43] But it doesn't have to be this way, he argues. We can alter that way of thinking. We can abandon the cultural directive to constantly compare ourselves with each other and turn to something more positive and beneficial for everyone:

> If you want to be important—wonderful. If you want to be recognized—wonderful. If you want to be great—wonderful. But recognize that he who is greatest among you shall be your servant. That's a new definition of greatness.
>
> ... By giving that definition of greatness, it means that everybody can be great, because everybody can serve. You don't have to have a college degree to serve. You don't have to make your subject and your verb agree to serve. You don't have to know about Plato and Aristotle to serve. You don't have to know Einstein's theory of relativity to serve. You don't have to know the second theory of thermodynamics in physics to serve. You only need a heart full of grace, a soul generated by love. And you can be that servant.[44]

We can decide to channel that instinct to be the best toward serving others. We can be the best at making the world a better place. We can put all that drive into service.

Interconnection

The conservationist John Muir said, in 1911, "When we try to pick out anything by itself, we find it hitched to everything else in the Universe."[45] Service isn't something done only for the benefit of others. Because we are interconnected, whatever impact I have on others will undoubtedly affect me too. As Dr. King put it, "We are caught in an inescapable network of mutuality, tied in a single garment of destiny. Whatever affects one directly, affects all indirectly."[46] Whether we can measure the benefits of service or not, whatever benefits I provide to one person, I provide to the whole, which includes myself, and thus improves the situation for everyone.

TYING IT ALL TOGETHER

With the Earth and all life that depends upon it hanging in the balance, living a simpler life has become an imperative. The world simply cannot sustain 7.8 billion people living the lifestyle of the average American or anything close to it. If we don't change our behaviors and tastes, driven by our underlying mindsets, we risk it all. But if we can shift our attitudes, minds, and behaviors toward living simply, with gratitude, with a purpose to serve, then not only can we avert planetary disaster, we may also find peace in our hearts and love in our communities. We may have more days filled with joy, ease, and happiness.

Mohandas K. Gandhi wrote about living a simple life back in 1905, far before the public was aware of the climate crisis, far before the age of mass consumerism had reached its current heights, back when the population of the world was only a fifth of what it is today. Still, even at that time, the same truth was evident. If we live a simpler life, we create more opportunity to help others and thus be happier.

> Happiness, the goal to which we all are striving is reached by endeavoring to make the lives of others happy, and if by renouncing the luxuries of life we can lighten the burdens of others . . . surely the simplification of our wants is a thing greatly to be desired! And so, if instead of supposing that we must become hermits and dwellers in caves in order to practice simplicity, we set about simplifying our affairs, each according to his own convictions and opportunity, much good will result and the simple life will at once be established.[47]

In 1967 Dr. Martin Luther King Jr. expressed similar concerns with our consumer-focused society.

> I am convinced that if we are to get on the right side of the world revolution, we as a nation must undergo a radical revolution of values. We must rapidly begin the shift from a "thing-oriented"

> society to a "person-oriented" society. When machines and computers, profit motives and property rights are considered more important than people, the giant triplets of racism, materialism, and militarism are incapable of being conquered.[48]

And of course, while it wasn't as pronounced in his day as it is now, environmental destruction would clearly fit on King's list of social ills. In fact, the orientation toward things that he described has continued and worsened to the present day, and is one of the largest factors driving environmental degradation.

It's not going to be easy to change the values of our culture. But nothing of great benefit comes easily. And it will be worth it. It will pay more dividends than we could ever imagine. A stable climate, a clean, healthy environment. More caring for each other, more sharing of resources, more concern for the well-being of others. Less consumption, less frenzy, less stress, less stuff. More mindfulness, more peace of mind, more love, more community. This is what the path forward looks like—for the climate, and also for our own well-being. How exactly we get there is up to each of us to think about and work out in our own lives, homes, communities, and in discussion with each other. But the goal of a world, of a society, of an America, that is more at peace with itself, more at ease, more grateful, and more in harmony with the natural world, is worth all the effort.

CHAPTER 10

ROLLING UP OUR SLEEVES

We ourselves are the pivotal human generation.

—DALAI LAMA XIV[1]

AS WE LEARNED FROM DR. KING, who emulated the techniques of Gandhi, who developed them based on the writings of Henry David Thoreau, the best way to create lasting change in society is not by besting your opponent, your oppressor, or adversary, but by appealing to their conscience. "One day we shall win freedom, but not only for ourselves. We shall so appeal to your heart and conscience that we shall win *you* in the process, and our victory will be a double victory," Dr. King wrote.[2]

Similarly, our goal is not to defeat those who would stall progress on climate change. Our goal is to find common ground that can bring those we have previously thought of as adversaries into the fight. Once we fully understand that we are all linked, that our fates are intertwined, then it will no longer be about pushing our agenda or fighting against theirs. It will be about finding a middle way, a shared path based on shared truths and values.

Doing that is easier than we might think. Because in the end we're not as different from each other as we may believe. All it takes is talking to each other more—and listening more—to find that out. It means spending less time on social media or staring at a screen and more time interacting with our community. The media and social media keep us divided by showing different audiences entirely different information. They do that on purpose. Their aim is to keep us afraid, outraged, disgusted, and on the edge of our seats, so that we keep tuning in. Don't believe it. Neither conservatives nor liberals are the gross stereotypical caricatures that they are depicted to be in the news and especially on social media.

At the end of the day, we're all just human beings trying to live good lives. To make progress as a country on climate or any other issue, we have to give each other the benefit of the doubt and have an open mind. We all must try to recognize that if we were in the other person's shoes, having lived their life, having been shaped by their experiences, we'd likely think a lot like they do—and vice versa.

Stay informed, but don't pay too much attention to that which you don't have control over. Focus on what you do have control over: your actions, your purchases, and your relationships with your neighbors and your community.

Start there.

Climate solutions won't come from the top down.

Nor will they come from trying to work all alone.

But when small groups come together at the *community level*, we can make changes that yield real benefits, which create new stories, which start trends, which shift popular opinion, which alters how the government and industries operate.

Remember, you're not alone in this fight. We're all concerned about climate change. So be the person who starts the conversation. You could even use your favorite anecdotes from this book. Perhaps it's how all the MGM hotels in Las Vegas are powered by solar and that by switching to renewables they're saving a ton of money, even after they had to pay an $87 million fine for leaving the municipal utility. Or perhaps it's the fact that the Yankees have decided to commit to the Paris Agreement. Or perhaps it's that five of the six companies with the highest valuation in the world are committed to 100 percent renewable energy. Or maybe it's that solar employs twice as many people in the power sector as the entire fossil fuel industry.

Maybe you're most excited by the fact that the cost of solar has dropped 70 percent in the past ten years, the fact that even the fossil fuel companies predict solar will be the dominant energy source within thirty years, or the fact that 100 percent clean energy will save Americans $1 trillion per year.

Or maybe you can't wait to share with someone the story of how Sally Bingham transformed her life by enrolling in college at the age of forty-five so she could go to seminary, become a priest, and lead a movement of green faith communities. Maybe you loved the description of how Republican mayor Dale Ross's city of Georgetown, Texas, is the second city in America to be 100 percent powered by renewable energy and how he fought to make it that way because it was the cost-effective thing to do. Perhaps you were impressed by the fact that the Department of Defense switched to solar energy on the front lines to save the lives of American soldiers.

Perhaps you didn't realize that in the aftermath of Hurricane Maria, Puerto Rico decided to completely reimagine its electrical grid and commit to 100 percent renewable energy with a resilient system of community-based mini-grids. Or that Solar Richmond and GRID Alternatives are teaching unemployed, at risk youth from historically disadvantaged communities to become solar installers.

I've tried in this book to highlight hopeful examples, the success stories in the climate fight—particularly those led by communities and particularly those that counter the divided partisan narrative. They may be small wins in the grand scheme of things, but lots of small wins amount to big victories. And if you read carefully, hopefully you can make out the patterns that led to these successes and see the road map that it provides for us moving forward.

TAKING ACTION

Throughout this book, we've explored several ways of thinking about climate change and talking about climate change, and we've learned about the positive momentum that we're building toward solutions. At the end of the day, though, we still have a heavy lift ahead of us. I want, then, to conclude this book with some thoughts on how we can all contribute to the transition to clean energy and sustainability in our local communities—where we have the most agency.

Climate change will not be solved by a silver bullet. Solar energy alone, for example, will not solve climate change. Creating an economy that's equitable and sustainable will require lots of technological adjustments, small and large, as well as new inventions, customs, laws, policies, and social norms. Furthermore, solutions must be localized, specialized, and appropriate for the culture in which they are meant to work. That means that we need everyone to contribute their expertise: no one single person, group of people, or country can solve the problem—but all together, we can. No matter your field of work, no matter where you live or what role you play in your home, workplace, or community, you and the people around you are interacting with nature and society and have insights into how to solve the problems associated with climate change. So that's where we start. Climate change solutions are not waiting for us at a fancy delegation of diplomats in a foreign country. They are at our kitchen table.

And remember, whatever you do, be it big or small, make sure it involves other people. Building community through climate solutions is the key to making them work.

Things Everyone Can Do

The list of climate actions I've compiled on the next few pages might seem like a lot at first glance. In order to not spook our Elephant remember to shrink the change to a manageable size. Instead of trying to do all twenty things on this list, start by picking one.

ENGAGE

- **Talk About Climate Change**

  Just talk about it. It all starts with conversation. Let people in your life know that this is important to you and ask them what it means to them. Especially talk with people who don't share your political views. Have the tough conversations. Get together with friends and family and talk about climate change. If you're feeling inspired, start a Climate Courage Circle and invite people to read and talk about stories and ideas about what solving climate change looks like. Visit climatecourage.

us to get started or to connect with a Climate Courage Circle in your area.

- **Be a Starter. Be a Joiner.**

  Start or join a climate initiative at work or in your neighborhood. Get together with your coworkers and brainstorm ways you can make your workplace or business practices more sustainable. Discuss with your neighbors or neighborhood association how you can make your community more sustainable, and then take action.

- **Help Change the Culture**

  Create art. Write poetry. Make music. Make films. Create sustainable businesses. Tell stories. Start conversations. Whatever you do, reach other people and stir them to action.

- **Vote and Get Involved in Local Politics**

  Look, I know our democracy isn't perfect. But using the vote to support climate leaders and policies can make an important difference, especially at the local level. Participating in our democracy matters and is a critical part of solving climate change. See what your town, city, and state can do about promoting clean energy and fighting climate change. Your voice matters. Make it heard. Join with some of the groups in this book to get involved in local campaigns.

- **Talk to Your Kids**

  If you're a parent, teach your kids the science of climate change, but also try not to scare them. Explain the best you can the political and social challenges to transitioning to renewable energy. But most importantly, empower them to do something. Go through this list and find things that kids can do on their own or with you to feel a sense of accomplishment.

ACCELERATE CLEAN ENERGY ADOPTION

- **Go Solar at Home**

  Save money on your electric bill. Increase the property value of your home. Support clean energy development. Reduce your environmental footprint. Get a quote from one of our trusted solar installer partners at re-volv.org. Are you a renter? See if you can subscribe to a community solar project nearby or sign up for green power from your utility.

- **Become a RE-volv Solar Ambassador**

  Can't go solar? Or you already have and are looking to help promote solar in other ways? Find a few friends willing to volunteer and look for nonprofits in your area to help go solar. We'll teach you the rest. Visit us at re-volv.org to learn more.

- **Divest from Fossil Fuels. Invest in Clean Energy.**

  Jim Cramer, former hedge fund manager and host of CNBC's *Mad Money*, declared in January 2020, "I'm done with fossil fuels. They're done. . . . We're starting to see divestment all over the world. . . . Big pension funds saying, 'Listen we're not gonna own them anymore.' We're in the death-knell phase. They're [the new] tobacco."[3] That's right. Thanks to 350.org and its many partners who launched a global fossil fuel divestment campaign, by 2020 they had already reached commitments from institutions around the globe to divest $14 trillion from fossil fuel holdings.[4] The trends are clear. Do yourself a favor now and divest from fossil fuels. Invest as much as you can in clean energy and climate solutions. Believe me, you'll be thankful you did.

CONSERVE RESOURCES AND REDUCE WASTE

- **Make Your Home More Energy Efficient**

  Get energy efficient lighting and appliances. Weatherize your home by sealing up your windows and doors and insulate your walls so the temperature-controlled air isn't always leaking out. Heat your space and water with an electric heat pump. Go with an electric stove. That way, when you go solar, you can power everything with the solar on your roof.

- **Reduce, Reuse, Consume Less, Embrace Enough**

  As the old expression goes, "use it up, wear it out, make do, or do without." And before you make a purchase, ask yourself if you really *need* this thing or if you merely *want* this thing. Try to resist the urge to consume more than you need to—for the sake of the planet and all its inhabitants.

- **Recycle, Compost, and Create a World Without Waste**

  In the groundbreaking book *Cradle to Cradle*, authors Michael Braungart and William McDonough explain how we can design and manufacture everything in our world to be used again and again by making as much as we can compostable and recycling the rest.[5] We can do our part to support that transition by recycling and composting as much as possible and demanding that product manufacturers design their products with their full life cycle in mind.

- **Decarbonize Your Transportation**

  You don't need to burn fossils to get from point A to B. If you're going short distances, ride your bike or walk. It's good exercise, fun, cheap, and great for the climate. For longer distances could you take public transportation, carpool, or rideshare? If you need your own car would an electric vehicle work for you? The cars are better built, they're cheaper to run thanks to no gas costs, and they're better for the environment—plus the ranges are getting better every day. Youth climate activist Greta Thunberg famously sailed across the Atlantic to make a point that there really aren't practical solutions yet for long distance sustainable transport. While those technologies are on the way, think about how you can reduce your footprint now. Flying, of course, has the highest carbon footprint of any means of transportation. Could you have a video conference instead of flying somewhere for work? How about taking a train? Could you choose going somewhere on vacation that doesn't involve flying?

EAT AND LIVE MINDFULLY

- **Create Community**

  Spending time with friends and family gives us the happiness and meaning we want in our lives and reduces our desire to fill the void with material consumption. Have your friends over for dinner. Enjoy good conversation and laughter with others more often.

- **Relax More**

  Take time for yourself. Read. Journal. Do yoga, meditate. Go for a run. Get a massage. Help promote the idea that we can slow down as a culture, be good to ourselves, and be more fulfilled as a result. You might consider Michael Pollan's suggestion to "observe the Sabbath. For one day a week, abstain completely from economic activity: no shopping, no driving, no electronics."[6] Or just embrace one day a week to rest and recharge however you see fit.

- **Eat Plants**

  Michael Pollan also famously suggested that we "eat food, not too much, mostly plants."[7] Turns out, that's not just good health advice, but it's also good for the planet. Eating a plant-based diet, or significantly reducing the amount of animal products we consume, can improve our health, reduce the suffering of animals, and be one of the most effective ways to reduce our impact on the planet.[8] The UN Food and Agriculture Organization estimates that livestock account for 14.5 percent of all anthropogenic greenhouse gas emissions.[9] And in terms of land use, livestock uses 83 percent of the world's farmland yet only provides 18 percent of the world's calories. Research shows that without meat and dairy consumption, global farmland could be reduced by 75 percent and still feed the world. Some of that land could be used to plant trees instead, thus pulling carbon out of the atmosphere, which is exactly what the UN Intergovernmental Panel on Climate Change calls for in its August 2019 report.[10] As Paul Hawken and his team write, in the book *Drawdown*, "Few climate solutions of this magnitude lie in the hands of individuals or are as close as the dinner plate."

- **Don't Waste Food**

  Looking for something easy to do to stop climate change? The *Drawdown* team has determined that one of top three things we can do to reduce carbon emissions is reducing food waste. As I discussed in chapter 9, about 40 percent of the food produced in the US is wasted.[11] Monitor how much you and your family consume and shop accordingly. Fill up your plate with smaller servings. Save what you can't finish for later. Or give it away. Just don't throw it away.

- **Garden**

  Grow some of your own food. It'll taste better, cost less, and dramatically reduce the carbon footprint of what you eat. In order to provide enough food for the soldiers overseas during World War II, Americans were encouraged to start "Victory Gardens." American families turned their front and backyards and vacant lots into gardens to grow food for themselves and their neighbors. In 1944, over twenty million victory gardens produced over eight million tons of food. These backyard gardens yielded over 40 percent of the country's vegetables.[12] Some call it the greatest voluntary movement the country has ever known.[13] That's exactly the type of citizen action we need to solve the climate crisis.

- **Buy Organic, Buy Local**

  When you are purchasing food, clothing, linens, cleaning products, personal-care products, and more, buy organic. It's better for your health, better for the workers, better for the soil, better for the water, better for the bees, better for the wildlife, better for the climate. And buying food and products sourced locally reduces how far they have to travel, and thus the associated carbon emissions, while also supporting your local economy. It's worth the extra few bucks.

- **Maybe One**

  If you're considering having kids or growing your family, read Bill McKibben's book *Maybe One*. In it he suggests considering having one child instead of two or three and describes the benefits of smaller families.

- **Stay Positive**

  I know thinking about climate change can be depressing. We're living through a very challenging moment in history. But the only way we're going to make it through is to keep our eyes on the prize, celebrate the victories big and small, and keep putting one foot in front of the other.

Did a suggestion from the list above jump out at you? Whatever it is, start there. Do one thing. See what you learn. See how it makes you feel. See the impact that it has. Share your results with others. Then come back here. Pick another item on the list. Do that. Come back and repeat the process again. Before you know it, you'll be crafting compelling new narratives around climate change while reducing your and your community's carbon impact significantly.

In addition to the list here, you can find out more ways to take action at climatecourage.us.

And remember, taking individual actions to reduce our impact on the planet is important, not for the sake of touting how green you are at social gatherings. And it's not because solving climate change is solely a matter of personal responsibility, as big polluters want us to believe, to remove the pressure on them to clean up their act. But the choices we make matter because of how they may influence others around us and ultimately shift our collective habits. What really matters is whether we get enough people to make different choices and adopt more sustainable behaviors, lifestyles, and practices often enough, so that they become the new norms and values embedded in the culture.

FINAL THOUGHTS

Would you agree with me that this place, this planet we're lucky enough to live on, is—pretty special? Look up at the sky. Look at the stars at night. For as far as we can see, for as far as our most advanced technology can see, we're the only ones around. Not a speck of life anywhere. Just cold, dark, empty space. And yet here, on this tiny blue dot in the sky is a cornucopia of beauty. This place is *magical*. This place has all the food and water and oxygen we need. This place is the perfect temperature for life to thrive, neither too hot nor too cold. And life has been thriving and evolving here for billions of years, filling every ecological niche in the most incredible ways imaginable. From the black and white stripes of the zebra, to the sprawling tentacles of the octopus, to the pink feathers of the flamingo, to the red and blue skin of tropical frogs, to the pink and purple and orange petals of flowers, to the mysterious translucent bodies of the creatures that live at the bottom of the ocean. To the taste of a delicious peach on your tongue. To the smell of a field after it rains. However this place came to be, it's gorgeous. I mean, it's really something. While the rest of the universe, as far as we know, is a desolate place, as you read these words, there are millions of people gathered together celebrating someone's birthday, or sharing a meal, or sharing a smile right now.

And from that place of appreciation for being alive, for getting to live in this paradise, can we find the courage to do what's right? Can we find the courage to put aside our self-interest, to put aside our differences, and to come together for the sake of this place we all love? Don't we owe it to life? To the universe? Don't we owe it to the land, to the seas, to the air, to all of the living things that share this place with us, for all they've done for us? Don't we know that it's not just about us?

As environmental activist and Buddhist scholar Joanna Macy puts it:

> To be alive in this beautiful, self-organizing universe—to participate in the dance of life with senses to perceive it, lungs that breathe it, organs that draw nourishment from it—is a wonder beyond words. And it is, moreover, an extraordinary privilege to be accorded a human life, with self-reflexive consciousness that brings awareness of our own actions and the ability to make choices. It lets us choose to take part in the healing of our world.[14]

If we get through this crisis successfully—if we successfully create a clean energy–powered world that no longer produces waste or pollution but rather creates abundance, that regenerates the earth rather than depletes it—then we can live in a world without scarcity and create harmony between humans and the rest of the world. And if we can achieve that—wow. In a world of clean, abundant energy, we can choose to create a world free of poverty. We can choose to create a world free of war. Because we'll have more than enough for everyone. That's the opportunity in front of us. Call me an idealist, but I believe that if we put our shoulder into it, we can get it done. And frankly, as an American, I welcome the challenge. Has there ever been a challenge we've put our full attention to as a nation that we weren't able to overcome? A goal we couldn't accomplish? If there's one thing that America has—in fact, you might call it our specialty—it's grit. We're a tenacious bunch. And now, my friends, is the time to get gritty. It's time to move past talking and get our hands dirty—in our communities—saving the world, this incredibly special place where you and I magically arrived. This paradise we're lucky enough to call home.

ACKNOWLEDGMENTS

FOR AS LONG AS I CAN REMEMBER I have wanted to write a book—this book. There have been so many people who have coached me, mentored me, advised me, and supported me along the way, and I won't have enough space to thank them all here. But I'll give it a shot. I want to acknowledge the following people for helping to make this book a reality:

Clint Wilder, my friend and RE-volv board member who listened to my concept for this book, thought it was a good idea, and connected me with his agent, Leah Spiro, not to mention all the advice on writing a book he's generously shared. Leah, who agreed to be my agent, worked with me to draft the proposal over many months and has been a tireless advocate for me and this book ever since. Amy Caldwell, my editor at Beacon Press, who took a chance on me as a first-time author and who patiently worked with me over many months to create this book in front of us. Amy helped shape the narrative and message of this book deeply, and I'm forever grateful for her guidance and wisdom. And all of the amazing staff at Beacon who have been true partners in this journey.

I want to thank all of my dear friends who read draft portions of the book, offered feedback, and advised me on the writing process, including Katharine Hayhoe, Clint Wilder, Paul Wapner, Sean Miller, Geoff Willard, Josh Turner, Emily Morris, Ambika Jain, Ari Natter, Gautham Rao, Danny Kennedy, and Bill McKibben.

I want acknowledge and thank all of the current and former RE-volv staff, board members, fellows, volunteers, funders, and supporters, as well as my mentors and advisors, who have worked tirelessly and given generously to help bring RE-volv's vision of community-based clean energy to life. It's through my work at RE-volv with these incredible people that I've learned about so much of what's captured in these pages.

I have tremendous gratitude for all the folks who tirelessly work for renewable energy and climate solutions and especially for those who gave me a chance to get professional experience and from whom I learned everything I know.

Mostly I want to thank all of the climate courage exemplars who are featured in the book who inspire me every day to push beyond what I perceive as limits as I see what you've been able to accomplish. Special thanks to the organizations and amazing people that partner so closely with RE-volv: Trisolaris, SunWork Renewable Energy Projects, GRID Alternatives, Solar United Neighbors, Vote Solar, the Sierra Club, Interfaith Power and Light, Green the Church, the Redford Center, the Climate Music Project, New Energy Nexus, the Local Clean Energy Alliance, the Asian Pacific Environmental Network, Communities for a Better Environment, the Clean Energy Leadership Institute, and many more.

I want to thank all of my teachers along the way who helped shape my thoughts, encourage creative pursuits, and hone my writing, especially Paul Wapner, Tony Seba, Holmes Hummel, Christos Kyrou, Ron Fisher, Mark Bergel, Roger Ahern, Mark Weissman, Don Knies, and Luci Hartmann.

I also want to thank all of the incredible environmental writers and thinkers whose works inspire me and have shaped my views. In particular I'm very grateful for the work of Mark Jacobson, whose research on technical clean energy solutions, and of Anthony Leiserowitz, whose research on the psychology of climate change communication, have helped shape this book.

Finally, I want to thank my mom, brother, and all my family, friends, and loved ones, who have all been so incredibly supportive this entire journey, and who have given me the space and encouragement to write, without which, this wouldn't have been possible.

NOTES

INTRODUCTION

1. Goldman Environmental Prize, "The Green Belt Movement: 40 Years of Impact," March 21, 2018, https://www.goldmanprize.org/blog/green-belt-movement-wangari-maathai.

2. "Last Time Carbon Dioxide Levels Were This High: 15 Million Years Ago, Scientists Report," *Science Daily*, UCLA, October 9, 2009, https://www.sciencedaily.com/releases/2009/10/091008152242.htm.

3. Jack Fitzpatrick, "Climatologist: Climate Science about as Certain as Theory of Gravity," *Morning Consult*, March 29, 2017, https://morningconsult.com/2017/03/29/climatologist-climate-science-certain-theory-gravity.

4. Martin Luther King Jr., *Strength to Love* (1963; Minneapolis: Fortress Press, 2010), 48.

5. Bill McKibben, "The Climate Science Is Clear: It's Now or Never to Avert Catastrophe," *Guardian*, November 20, 2019, https://www.theguardian.com/commentisfree/2019/nov/20/climate-crisis-its-now-or-never-to-avert-catastrophe.

6. McKibben, "The Climate Science Is Clear."

CHAPTER 1: REFRAMING THE NARRATIVE

1. Kate Marvel, "We Need Courage, Not Hope, to Face Climate Change," *On Being*, March 1, 2018, https://onbeing.org/blog/kate-marvel-we-need-courage-not-hope-to-face-climate-change.

2. Bob Inglis, "A Conservative Who Believes That Climate Change Is Real," interview by Roger Cohn, *Yale Environment 360*, February 14, 2013, https://e360.yale.edu/features/interview_bob_inglis_conservative_who_believes_climate_change_is_real.

3. Cecile Andrews and Wanda Urbanska, *Less Is More: Embracing Simplicity for a Healthy Planet, a Caring Economy, and Lasting Happiness* (Gabriola Island, BC: New Society, 2009), xiv.

4. Piers Steel, *The Procrastination Equation: How to Stop Putting Things Off and Start Getting Stuff Done* (New York: HarperCollins, 2011), 67.

CHAPTER 2: THE PSYCHOLOGY OF CLIMATE CHANGE

1. Bartholomew, Ecumenical Patriarch of Constantinople, *On Earth as in Heaven: Ecological Vision and Initiatives of Ecumenical Patriarch Bartholomew* (New York: Fordham University Press, 2011), 223.

2. Henry Fountain, "Climate Change Is Accelerating, Bringing World 'Dangerously Close' to Irreversible Change," *New York Times*, December 4, 2019, https://www.nytimes.com/2019/12/04/climate/climate-change-acceleration.html?

3. "No One Should Want Their Children to Live in This 'Bleak' Future," editorial, *Washington Post*, December 1, 2019, https://www.washingtonpost.com/opinions/no-one-should-want-their-children-to-live-in-this-bleak-future/2019/12/01/70771a84-1159-11ea-b0fc-62cc38411ebb_story.html.

4. David Wallace-Wells, *The Uninhabitable Earth: Life after Warming* (New York: Tim Duggan Books, 2019).

5. Jonathan Franzen, "What If We Stopped Pretending?," *New Yorker*, September 8, 2019, https://www.newyorker.com/culture/cultural-comment/what-if-we-stopped-pretending.

6. Mark Z. Jacobson, "No, #globalwarming IS a solvable problem and 61 countries have already passed laws to partly get there," Twitter, September 8, 2019, 10:04 p.m., https://twitter.com/mzjacobson/status/1170895604685262848.

7. Kate Marvel, "Shut Up, Franzen," *Scientific American*, September 11, 2019, https://blogs.scientificamerican.com/hot-planet/shut-up-franzen.

8. Daniel Goleman, *Focus: The Hidden Driver of Excellence* (New York: HarperCollins, 2013), 148.

9. George Marshall, *Don't Even Think About It: Why Our Brains Are Wired to Ignore Climate Change* (New York: Bloomsbury, 2015), 56.

10. Marshall, *Don't Even Think About It*, 56–7.

11. Marshall, *Don't Even Think About It*, 57.

12. Marshall, *Don't Even Think About It*, 58.

13. Paulina Dedaj, "Bahamas Say 2,500 People Missing after Hurricane Dorian, Death Toll 'Expected to Significantly Increase,'" Fox News, September 12, 2019, https://www.foxnews.com/world/bahamas-missing-hurricane-dorian-increase-death-toll.

14. Brady Dennis and Chris Mooney, "Wildfires, Hurricanes and Other Extreme Weather Cost the Nation 247 Lives, Nearly $100 Billion in Damage During 2018," *Washington Post*, February 6, 2019, https://www.washingtonpost.com/climate-environment/2019/02/06/wildfires-hurricanes-other-extreme-weather-cost-nation-lives-nearly-billion-damage-during.

15. Anthony Leiserowitz et al., *Climate Change in the American Mind: December 2018*, Yale University and George Mason University (New Haven, CT: Yale Program on Climate Change Communication), https://climatecommunication.yale.edu/wp-content/uploads/2019/01/Climate-Change-American-Mind-December-2018.pdf, 3.

16. Chip Heath and Dan Heath, *Switch: How to Change Things When Change Is Hard* (New York: Crown, 2010), 5.

17. Simon Sinek, *Start with Why: How Great Leaders Inspire Everyone to Take Action* (New York: Portfolio, 2009), 59–60.
18. Rand Swenson, "Review of Clinical and Functional Neuroscience," Dartmouth Medical School, 2006, https://www.dartmouth.edu/fflrswenson/NeuroSci/chapter_9.html.
19. Sinek, *Start with Why*, 56.
20. Steel, *The Procrastination Equation*, 73.
21. Heath and Heath, *Switch*, 7–8.
22. Daniel Kahneman, *Thinking, Fast and Slow* (New York: Farrar, Straus and Giroux, 2013), 31.
23. Heath and Heath, *Switch*, 112–13.
24. Sinek, *Start with Why*, 56.
25. Heath and Heath, *Switch*, 106.
26. Heath and Heath, *Switch*, 147.
27. Sinek, *Start with Why*, 129.
28. Sinek, *Start with Why*, 128.
29. Goleman, *Focus*, 151.
30. Chip Heath, "An Interview with Author Chip Heath about Making Environmental Messages Sticky," interview by *Grist* staff, *Grist*, March 13, 2007, https://grist.org/article/heath.
31. Goleman, *Focus*, 153.
32. Charles Duhigg, *The Power of Habit: Why We Do What We Do in Life and Business* (New York: Random House, 2014), 277.
33. Heath and Heath, *Switch*, 122.
34. Heath and Heath, *Switch*, 122.
35. Heath and Heath, *Switch*, 123.
36. Heath and Heath, *Switch*, 129.
37. Heath and Heath, *Switch*, 28.
38. Heath and Heath, *Switch*, 31.
39. Heath and Heath, *Switch*, 129.
40. Heath and Heath, *Switch*, 133–34.
41. Heath and Heath, *Switch*, 134.
42. Heath and Heath, *Switch*, 143.
43. Marshall, *Don't Even Think About It*, 215.
44. Duhigg, *The Power of Habit*, 89.
45. Sinek, *Start with Why*, 55.
46. Sinek, *Start with Why*, 129–30.
47. Duhigg, *The Power of Habit*, 217.
48. Duhigg, *The Power of Habit*, 242.
49. Duhigg, *The Power of Habit*, 243.
50. Heath and Heath, *Switch*, 153
51. Henry Wismayer, "Liberals: Please Chill Out," *Medium*, June 15, 2018, https://medium.com/s/jeremiad/liberals-please-chill-out-7f7309e4d364.

CHAPTER 3: CLIMATE CONFUSION

1. Angela Duckworth, *Grit: The Power of Passion and Perseverance* (New York: Scribner, 2016), 178.

2. Philip Shabecoff, "Global Warming Has Begun, Expert Tells Senate," *New York Times*, June 24, 1988, https://www.nytimes.com/1988/06/24/us/global-warming-has-begun-expert-tells-senate.html.

3. Scott Waldman and Benjamin Hulac, "This Is When the GOP Turned Away from Climate Policy," *E&E News*, December 5, 2018, https://www.eenews.net/stories/1060108785.

4. Waldman and Hulac, "This Is When the GOP Turned Away from Climate Policy."

5. George H. W. Bush, "1992 Rio Summit," C-SPAN, June 13, 1992, user clip video, 1:29, https://www.c-span.org/video/?c4792507/user-clip-1992-rio-summit.

6. "Theodore Roosevelt and Conservation," National Park Service, https://www.nps.gov/thro/learn/historyculture/theodore-roosevelt-and-conservation.htm.

7. John Schwartz, "The 'Profoundly Radical' Messages of Earth Day's First Organizer," *New York Times*, April 20, 2020, https://www.nytimes.com/2020/04/20/climate/denis-hayes-earth-day-organizer.html.

8. 8. "The Origins of EPA," Environmental Protection Agency, https://www.epa.gov/history/origins-epa; "National Environmental Policy Act," Department of Energy, https://ceq.doe.gov; "Evolution of the Clean Air Act," EPA, https://www.epa.gov/clean-air-act-overview/evolution-clean-air-act, accessed May 4, 2020; "Summary of the Clean Water Act," EPA, https://www.epa.gov/laws-regulations/summary-clean-water-act, accessed May 4, 2020.

9. "40th Anniversary of the Clean Air Act," Environmental Protection Agency, https://www.epa.gov/clean-air-act-overview/40th-anniversary-clean-air-act.

10. "President Bush's Response to Global Warming," *PBS News Hour*, June 11, 2001, https://www.pbs.org/newshour/show/president-bushs-response-to-global-warming.

11. Katie Couric, "The Candidates on Climate Change," *CBS Evening News*, December 11, 2007, https://www.cbsnews.com/news/the-candidates-on-climate-change.

12. Office for Commonwealth Development, *Massachusetts Climate Protection Plan*, Commonwealth of Massachusetts, 2004, https://www.documentcloud.org/documents/499572-romney-massachusetts-climate-action-plan-2004.html.

13. Alexander Bolton, "Romney Helps GOP Look for New Path on Climate Change," *Hill*, March 24, 2019, https://thehill.com/homenews/senate/435328-romney-helps-gop-look-for-new-path-on-climate-change.

14. Ben Adler and Rebecca Leber, "Donald Trump Once Backed Urgent Climate Action. Wait, What?," *Grist*, June 8, 2016, https://grist.org/politics/donald-trump-climate-action-new-york-times.

15. Donald Trump, "Donald Trump's *New York Times* Interview: Full Transcript," *New York Times*, November 23, 2016, https://www.nytimes.com/2016/11/23/us/politics/trump-new-york-times-interview-transcript.html?module=inline.

16. James Inhofe, "*The Rachel Maddow Show*, Transcript 03/15/12," interview by Rachel Maddow, MSNBC, March 15, 2012, http://www.msnbc.com/transcripts/rachel-maddow-show/2012-03-15.

17. Anthony Leiserowitz, email message to author, February 15, 2012.

18. Justin Rolfe-Redding et al., "Republicans and Climate Change: An Audience Analysis of Predictors for Belief and Policy Preferences," Center for Climate Change Communication, George Mason University (2012): 30–31, SSRN: https://ssrn.com/abstract=2026002 or http://dx.doi.org/10.2139/ssrn.2026002.

19. Rolfe-Redding et al., "Republicans and Climate Change," 31.

20. Juliet Eilperin, Josh Dawsey, and Brady Dennis, "White House to Set Up Panel to Counter Climate Change Consensus, Officials Say," *Washington Post*, February 24, 2019, https://www.washingtonpost.com/national/health-science/white-house-to-select-federal-scientists-to-reassess-government-climate-findings-sources-say/2019/02/24/49cd0a84-37dd-11e9-af5b-b51b7ff322e9_story.html.

21. Coral Davenport and Kendra Pierre-Louis, "U.S. Climate Report Warns of Damaged Environment and Shrinking Economy," *New York Times*, November 23, 2018, https://www.nytimes.com/2018/11/23/climate/us-climate-report.html.

22. "Trump on Climate Change Report: 'I Don't Believe It,'" BBC News, November 26, 2018, video, 0:40, https://www.bbc.com/news/world-us-canada-46351940.

23. Jordain Carney, "McConnell: 'I Do' Believe in Human-Caused Climate Change," *Hill*, March, 26, 2019, https://thehill.com/homenews/senate/435904-mcconnell-i-do-believe-in-human-caused-climate-change.

24. Rebecca Beitsch, "Bipartisan Senate Climate Caucus Grows by Six Members," *Hill*, November 6, 2019, https://thehill.com/policy/energy-environment/469242-bipartisan-senate-climate-caucus-grows-by-six-members.

25. Darren Samuelsohn, "Meet Lindsey Graham, the Next GOP Maverick on Climate Change," *New York Times*, October 13, 2009, https://archive.nytimes.com/www.nytimes.com/cwire/2009/10/13/13climatewire-meet-lindsey-graham-the-next-gop-maverick-on-13485.html?scp=2&sq=climatewire&st=cse.

26. John Boehner, interview by Steve Inskeep, *Morning Edition*, NPR News, January 22, 2009, https://www.npr.org/transcripts/99653076?storyId=99653076?storyId=99653076.

27. John Boehner, interview by Wolf Blitzer, *The Situation Room*, CNN, July 15, 2008, http://transcripts.cnn.com/TRANSCRIPTS/0807/15/sitroom.01.html.

28. Zachary Coile, "Pelosi, Gingrich Team Up for Global Warming TV Ad," *SFGate*, April 18, 2018, https://blog.sfgate.com/politics/2008/04/18/pelosi-gingrich-team-up-for-global-warming-tv-ad.

29. Newt Gingrich, "Interviews: Newt Gingrich," *Frontline: Politics*, April 24, 2007, https://www.pbs.org/wgbh/pages/frontline/hotpolitics/interviews/gingrich.html.

30. Newt Gingrich, "Interview: Newt Gingrich," interview by Andrew Revkin, *New York Times*, November 12, 2007, video, 5:25, https://www.nytimes.com/video/science/1194817121604/interview-newt-gingrich.html.

31. Carla Marinucci, "Giuliani Says He's '100 Percent Committed' to Running for President," *SFGate*, February 12, 2007, https://www.sfgate.com/politics/article/Giuliani-says-he-s-100-percent-committed-to-2649401.php.

32. "When Republican Sen. [and Republican presidential nominee] Bob Dole from Kansas first heard about global warming 30 to 40 years ago, he thought it was an issue that deserved attention." Gary Burt, "Up to Democrats to Responsibly Fix Climate," op-ed, *Duluth News Tribune*, November 6, 2018, https://www.duluthnewstribune.com/opinion/4525293-readers-view-democrats-responsibly-fix-climate.

33. Center for Responsive Politics, "Oil & Gas: Long-Term Contribution Trends," OpenSecrets.org, https://www.opensecrets.org/industries/totals.php?cycle=2018&ind=E01, accessed March 12, 2020.

34. Center for Responsive Politics, "Electric Utilities: Long-Term Contribution Trends," OpenSecrets.org, https://www.opensecrets.org/industries/totals.php?ind=E08++, accessed March 12, 2020.

35. Center for Responsive Politics, "Coal Mining: Long-Term Contribution Trends, " OpenSecrets.org, https://www.opensecrets.org/industries/totals.php?ind=E1210, accessed March 12, 2020.

36. Center for Responsive Politics, "Oil & Gas: Top Recipients," Open Secrets.org, https://www.opensecrets.org/industries/recips.php?ind=E01&recipdetail=A&sortorder=U&mem=Y&cycle=2016, accessed March 12, 2020.

37. Benjamin Hulac, "Tobacco and Oil Industries Used Same Researchers to Sway Public," *E&E News* for *Scientific American*, July 20, 2016, https://www.scientificamerican.com/article/tobacco-and-oil-industries-used-same-researchers-to-sway-public1.

38. Roger Revelle et al., "Appendix Y4: Atmospheric Carbon Dioxide," in *Restoring the Quality of Our Environment*, Report of the Environmental Pollution Panel, President's Science Advisory Committee (Washington, DC: White House, 1965), 126–27, https://assets.documentcloud.org/documents/3227654/PSAC-1965-Restoring-the-Quality-of-Our-Environment.pdf.

39. Sharon Kelly, "'Time is Running Out,' American Petroleum Institute Chief Said in 1965 Speech on Climate Change," *Desmog*, November 20, 2018, https://www.desmogblog.com/2018/11/20/american-petroleum-institute-1965-speech-climate-change-oil-gas.

40. E. Robinson, and R. C. Robbins, *Sources, Abundance, and Fate of Atmospheric Pollutants* (Menlo Park, CA: Stanford Research Institute, 1968). Excerpts from the report (pages 108–110, 112) can be viewed in the Center for International Environmental Law timeline *Smoke & Fumes*, https://www.smokeandfumes.org/documents/document16.

41. Neela Banerjee, "How Big Oil Lost Control of Its Climate Misinformation Machine," *InsideClimate News*, December 22, 2017, https://insideclimatenews.org/news/22122017/big-oil-heartland-climate-science-misinformation-campaign-koch-api-trump-infographic.

42. Bill McKibben, "How Extreme Weather Is Shrinking the Planet," *New Yorker*, November 16, 2018, https://www.newyorker.com/magazine/2018/11/26/how-extreme-weather-is-shrinking-the-planet.

43. McKibben, "How Extreme Weather Is Shrinking the Planet."

44. Joe Romm, "Republicans Turn to Frank Luntz's Playbook Instead of Developing a Real Climate Plan," *ThinkProgress*, March 27, 2019. https://thinkprogress.org/republican-green-new-deal-fossil-fuels-c5585c544718.

45. McKibben, "How Extreme Weather Is Shrinking the Planet."

46. Hulac, "Tobacco and Oil Industries Used Same Researchers to Sway Public."

47. Hulac, "Tobacco and Oil Industries Used Same Researchers to Sway Public."

48. Banerjee, "How Big Oil Lost Control of Its Climate Misinformation Machine."

49. Peter C. Frumhoff and Naomi Oreskes, "Fossil Fuel Firms Are Still Bankrolling Climate Denial Lobby Groups," *Guardian*, March 25, 2015, https://www.theguardian.com/environment/2015/mar/25/fossil-fuel-firms-are-still-bankrolling-climate-denial-lobby-groups.

50. These datapoints are highlighted in Sandra Laville, "Top Oil Firms Spending Millions Lobbying to Block Climate Change Policies, Says Report," *Guardian*, March 21, 2019, https://www.theguardian.com/business/2019/mar/22/top-oil-firms-spending-millions-lobbying-to-block-climate-change-policies-says-report.

51. Publicly available records. For all companies' 2018 revenues, except Chevron, see Adam Muspratt, "The Top 10 Oil & Gas Companies in the World: 2019," Oil & Gas IQ, May 1, 2019, https://www.oilandgasiq.com/strategy-management-and-information/articles/oil-and-gas-companies. For Chevron, see "Chevron Net Income 2006-2019/CVX," Macrotrends, https://www.macrotrends.net/stocks/charts/CVX/chevron/revenue, accessed March 12, 2020.

52. Banerjee, "How Big Oil Lost Control of Its Climate Misinformation Machine."

53. Rolfe-Redding et al., "Republicans and Climate Change," 5.

54. Jonathan Mahler and Jim Rutenberg, "Planet Fox: How Rupert Murdoch's Empire of Influence Remade the World. Part 3: The New Fox

Weapon," *New York Times Magazine*, April 3, 2019, https://www.nytimes.com/interactive/2019/04/03/magazine/new-fox-corporation-disney-deal.html.

55. Scientific Consensus: Earth's Climate Is Warming," Global Climate Change, NASA, https://climate.nasa.gov/scientific-consensus, accessed May 10, 2020.

56. "Scientific Consensus: Earth's Climate Is Warming," Global Climate Change, NASA, accessed May 10, 2020; Dana Nuccitelli, "Conservative Media Outlets Found Guilty of Biased Global Warming Coverage," *Guardian*, October 11, 2013, https://www.theguardian.com/environment/climate-consensus-97-per-cent/2013/oct/11/climate-change-political-media-ipcc-coverage.

57. Evlondo Cooper, "ABC, CBS, and NBC Completely Failed to Mention Climate Change in Coverage of Major Midwest Floods," *Media Matters*, March 29, 2019, https://www.mediamatters.org/blog/2019/03/29/ABC-CBS-and-NBC-completely-failed-to-mention-climate-change-in-coverage-of-major-Midwest-f/223270.

58. Cooper, "ABC, CBS, and NBC Completely Failed to Mention Climate Change in Coverage of Major Midwest Floods."

CHAPTER 4: CONSERVATIVES FORGING A PATH

1. Abraham Lincoln, "In the First Debate with Douglas," *The World's Famous Orations*, 1858, https://www.bartleby.com/268/9/23.html.

2. Leiserowitz et al., *Climate Change in the American Mind*, 4.

3. Anthony Leiserowitz et al., *Politics & Global Warming, November 2019*, Yale University and George Mason University (New Haven, CT: Yale Program on Climate Change Communication, 2020), 4-6, https://climatecommunication.yale.edu/wp-content/uploads/2020/01/politics-global-warming-november-2019b.pdf

4. Anthony Leiserowitz et al., *Energy in the American Mind: December 2018*, Yale University and George Mason University (New Haven, CT: Yale Program on Climate Change Communication, 2018), 4–5, https://climatecommunication.yale.edu/wp-content/uploads/2019/02/Energy-American-Mind-December-2018.pdf.

5. Leiserowitz et al., *Energy in the American Mind*, 4.

6. Arnold Schwarzenegger, "Schwarzenegger: No Country More Welcoming Than the USA," CNN.com, August 31, 2004, https://www.cnn.com/2004/ALLPOLITICS/08/31/gop.schwarzenegger.transcript.

7. Arnold Schwarzenegger, "Trump Can't Erase a Decade of Clean Air Progress with a Sharpie," *Washington Post*, September 8, 2019, https://www.washingtonpost.com/opinions/trump-cant-erase-a-decade-of-clean-air-progress-with-a-sharpie/2019/09/08/8d6393de-d248-11e9-86ac-0f250cc91758_story.html.

8. Mary Crane, "The Governator's Shade of Green," *Forbes*, April 12, 2007, https://www.forbes.com/2007/04/12/schwarzenegger-eco-speech-biz-man-cx_mc_0412bizeco.html#7a582de53661.

9. Colin Woodard, "America's First All-Renewable-Energy City," *Politico*, November 17, 2016, https://www.politico.com/magazine/story/2016/11/burlington-what-works-green-energy-214463.

10. Dan Solomon, "Is a Texas Town the Future of Renewable Energy?," *Smithsonian Magazine*, April 2018, https://www.smithsonianmag.com/innovation/texas-town-future-renewable-energy-180968410.

11. Solomon, "Is a Texas Town the Future of Renewable Energy?"

12. Solomon, "Is a Texas Town the Future of Renewable Energy?"

13. Jonathan Tilove, "How Georgetown's GOP Mayor Became a Hero to Climate Change Evangelists," *Statesman*, October 23, 2017, https://www.statesman.com/article/20171023/news/310239786.

14. Dale Ross, "Mayor: Why My Texas Town Ditched Fossil Fuel," *Time*, March 27, 2015, https://time.com/3761952/georgetown-texas-fossil-fuel-renewable-energy.

15. Ross, "Mayor."

16. Ross, "Mayor."

17. Ross, "Mayor."

18. Solomon, "Is a Texas Town the Future of Renewable Energy?"

19. Debbie Dooley, "This Green Tea Party Star Is Fighting for Solar," interviewed by Kari Lydersen, *Grist*, September 14, 2014, https://grist.org/politics/this-green-tea-party-star-is-fighting-for-solar.

20. Rob Wile, "A Group of Snipers Shot Up a Silicon Valley Power Station for 19 Minutes Last Year Before Slipping into the Night," *Business Insider*, February 5, 2014, https://www.businessinsider.com/silicon-valley-power-station-sniper-attack-2014–2.

21. Carolyn Kormann, "Greening the Tea Party," *New Yorker*, February 17, 2015, https://www.newyorker.com/tech/annals-of-technology/green-tea-party-solar.

22. Dooley, "This Green Tea Party Star Is Fighting for Solar."

23. Suzanne Goldenberg and Ed Pilkington, "ALEC Calls for Penalties on 'Freerider' Homeowners in Assault on Clean Energy," *Guardian*, December 4, 2013, https://www.theguardian.com/world/2013/dec/04/alec-freerider-homeowners-assault-clean-energy.

24. Kormann, "Greening the Tea Party."

25. Dooley, "This Green Tea Party Star Is Fighting for Solar."

26. Dooley, "This Green Tea Party Star Is Fighting for Solar."

27. Dooley, "This Green Tea Party Star Is Fighting for Solar."

28. Tim Dickinson, "Study: US Fossil Fuel Subsidies Exceed Pentagon Spending," *Rolling Stone*, May 8, 2019, https://www.rollingstone.com/politics/politics-news/fossil-fuel-subsidies-pentagon-spending-imf-report-833035.

29. Dooley, "This Green Tea Party Star Is Fighting for Solar."

30. Matteen Mokalla, "I'm a Tea Party Conservative. Here's How to Win Over Republicans on Renewable Energy," *Vox*, April 18, 2017, https://www.vox.com/videos/2017/4/18/15339266/debbie-dooley-tea-party-conservative

-republicans-renewable-energy. All of the quotes in this paragraph come from this article.

31. James Rainey, “Bob Inglis, a Republican Believer in Climate Change, Is Out to Convert His Party,” NBC News, September 30, 2018, https://www.nbcnews.com/news/us-news/bob-inglis-republican-believer-climate-change-out-convert-his-party-n912066.

32. Rainey, “Bob Inglis, a Republican Believer in Climate Change.”

33. Rainey, “Bob Inglis, a Republican Believer in Climate Change.”

34. Bob Inglis, “American Bipartisan Politics Can Be Saved—Here’s How,” TEDxBeaconStreet, November 2017, video, 10:48, https://www.ted.com/talks/bob_inglis_american_bipartisan_politics_can_be_saved_here_s_how/transcript?language=en.

35. Bob Inglis, “I Wasn’t Always So Courageous on Climate,” excerpt from John F. Kennedy Profile in Courage Award acceptance speech, May 3, 2015, *Slate*, June 18, 2018, video, 1:47, https://www.youtube.com/watch?v=la9jDbhs_-E.

36. “Our Principles,” RepublicEn, https://www.republicen.org/about_us/principles.

37. “Our Principles,” RepublicEn.

38. Sean Hannity, “Does the Democrats’ ‘Green New Deal’ Weaponize Climate Change?,” *Hannity*, February 7, 2019, https://www.foxnews.com/transcript/does-the-democrats-green-new-deal-weaponize-climate-change.

39. “The Green New Deal Is Better Than Our Climate Nightmare,” editorial, *New York Times*, February 23, 2019, https://www.nytimes.com/2019/02/23/opinion/green-new-deal-climate-democrats.html.

40. Citizens’ Climate Lobby, “CCL Praises Formation of Bipartisan House Climate Solutions Caucus,” news release, February 3, 2016, https://citizensclimatelobby.org/ccl-praises-formation-of-bipartisan-house-climate-solutions-caucus.

41. Natasha Geiling, “A Caucus and Bull Story,” *Grist*, October 23, 2018, https://grist.org/article/the-midterms-could-spell-the-end-of-this-bipartisan-climate-caucus.

42. Erin Mundahl, “2018 Decimates Bipartisan Climate Solutions Caucus,” *Inside Sources*, November 9, 2018, https://www.insidesources.com/2018-decimates-climate-solutions-caucus-is-this-the-end-of-bipartisan-climate-action.

43. Rebecca Beitsch,“Bipartisan Senate Climate Caucus Grows by Six Members,” *Hill*, November 6, 2019, https://thehill.com/policy/energy-environment/469242-bipartisan-senate-climate-caucus-grows-by-six-members.

44. “Majorities See Government Efforts to Protect the Environment as Insufficient,” Pew Research Center, May 14, 2018, https://www.pewresearch.org/science/2018/05/14/majorities-see-government-efforts-to-protect-the-environment-as-insufficient.

45. Amanda Paulson, "Why These Young Republicans See Hope in Climate Action," *Christian Science Monitor*, June 28, 2018, https://www.csmonitor.com/Environment/2018/0628/Why-these-young-Republicans-see-hope-in-climate-action.

46. Young Conservatives for Energy Reform, https://yc4er.org, accessed February 10, 2020.

47. "About Us," Young Conservatives for Energy Reform, http://yc4er.org/about-us, accessed February 10, 2020.

48. "About Us," Young Conservatives for Energy Reform.

49. Robinson Meyer, "They're Here to Fix Climate Change! They're College Republicans," *Atlantic*, February 28, 2018, https://www.theatlantic.com/science/archive/2018/02/college-republicans-carbon-climate-change-plan/554465.

50. Meyer, "They're Here to Fix Climate Change!"

51. Paulson, "Why These Young Republicans See Hope."

52. Students for Climate Dividends, https://www.s4cd.org, accessed February 10, 2020.

53. Meyer, "They're Here to Fix Climate Change!"

54. "Founding Member Statements," Climate Leadership Council, https://clcouncil.org/statements, accessed March 12, 2020.

55. "Founding Members," Climate Leadership Council, http://clcouncil.org/founding-members, accessed March 12, 2020.

56. "Strategic Partners," Climate Leadership Council, http://clcouncil.org/strategic-partners, accessed March 12, 2020.

57. *National Survey Results on the Baker-Shultz Carbon Dividends Plan*, Yale University and George Mason University (New Haven, CT: Yale Program on Climate Change Communication, 2018), https://www.clcouncil.org/media/YaleGMU-Poll-October-2018.pdf.

58. Abel Gustafson et al., *The Green New Deal Has Strong Bipartisan Support*, Yale University and George Mason University (New Haven, CT: Yale Program on Climate Change Communication, 2018), http://climatecommunication.yale.edu/publications/the-green-new-deal-has-strong-bipartisan-support.

59. Marianne Lavelle, "Green New Deal vs. Carbon Tax: A Clash of 2 Worldviews, Both Seeking Climate Action," *InsideClimate News*, March 4, 2019, https://insideclimatenews.org/news/04032019/green-new-deal-carbon-tax-compromise-climate-policy-congress-ocasio-cortez-sunrise-ccl-economists.

CHAPTER 5: JOBS, JOBS, JOBS

1. Ralph Waldo Emerson, *Self-Reliance and Other Essays* (New York: Dover, 1993), 2.

2. Leta Dickinson, "Renewable Energy Outpaced Coal in April for the First Time Ever," *Grist*, May 2, 2019, https://grist.org/article/renewable-energy-outpaced-coal-in-april-for-the-first-time-ever.

3. "The Birth of the IBM PC," IBM, https://www.ibm.com/ibm/history/exhibits/pc25/pc25_birth.html.

4. Tibi Puiu, "Your Smartphone Is Millions of Times More Powerful Than All of NASA's Combined Computing in 1969," *ZME Science*, October 13, 2015, https://www.zmescience.com/research/technology/smartphone-power-compared-to-apollo-432.

5. John Weaver, "California Residential Solar Power Headed toward $1/W and 2.5¢/kWh," *PV Magazine*, May 14, 2018, https://pv-magazine-usa.com/2018/05/14/california-residential-solar-power-headed-to-1-12-w-2-5%C2%A2-kwh.

6. Solar Energy Industries Association, "Solar Industry Research Data," 2019, https://www.seia.org/solar-industry-research-data.

7. Jason Deign, "Why PV Costs Have Fallen So Far—and Will Fall Further," *Greentech Media*, December 14, 2018, https://www.greentechmedia.com/articles/read/why-pv-costs-have-fallen-so-far-and-will-fall-further#gs.3kzzrg.

8. David R. Baker, "Gas Plants Will Get Crushed by Wind, Solar by 2035, Study Says," *Bloomberg*, September 9, 2019, https://www.bloomberg.com/news/articles/2019-09-09/gas-plants-will-get-crushed-by-wind-solar-by-2035-study-says. Nathaniel Bullard, "The New Math for Investors in a World of Cheap Energy," *Bloomberg*, May 7, 2020, https://www.bloomberg.com/news/articles/2020-05-07/the-new-investor-math-in-the-wake-of-the-oil-market-crash.

9. Mark Z. Jacobson, "Evaluation of Nuclear Power as a Proposed Solution to Global Warming, Air Pollution, and Energy Security," in *100% Clean, Renewable Energy and Storage for Everything* (Cambridge, UK: Cambridge University Press, 2020), https://web.stanford.edu/group/efmh/jacobson/Articles/I/NuclearVsWWS.pdf.

10. McKinsey and Company, *Global Energy Perspective 2019: Reference Case*, Energy Insights, January 2019, https://www.mckinsey.com/ffl/media/McKinsey/Industries/Oil%20and%20Gas/Our%20Insights/Global%20Energy%20Perspective%202019/McKinsey-Energy-Insights-Global-Energy-Perspective-2019_Reference-Case-Summary.ashx.

11. Solar Energy Industries Association and Wood Mackenzie Power & Renewables, *Solar Market Insight Report: 2018 Year in Review*, March 13, 2019, https://www.seia.org/research-resources/solar-market-insight-report-2018-year-review.

12. David Roberts, "Here's What It Would Take for the US to Run on 100% Renewable Energy," *Vox*, May 3, 2016, https://www.vox.com/2015/6/9/8748081/us-100-percent-renewable-energy.

13. Roberts, "Here's What It Would Take for the US to Run on 100% Renewable Energy."

14. Mark Delucchi and Mark Z. Jacobson, "Why the Green New Deal Cuts Consumer Energy Costs & Unemployment," *CleanTechnica*, March 9,

2019, https://cleantechnica.com/2019/03/09/why-the-green-new-deal-cuts-consumer-energy-costs-unemployment.

15. Delucchi and Jacobson, "Why The Green New Deal Cuts Consumer Energy Costs."

16. Carolyn Fortuna, "The Shell Sky Scenario: Our Need to Critically Assess Sponsored Corporate Content," *CleanTechnica*, November 30, 2018, https://cleantechnica.com/2018/11/30/the-shell-sky-scenario-our-need-to-critically-assess-sponsored-corporate-content.

17. E2 and Clean Jobs Count, *Clean Jobs America*, March 2019, https://www.e2.org/wp-content/uploads/2019/04/E2-2019-Clean-Jobs-America.pdf.

18. US Energy Information Administration, *U.S. Energy Facts Explained*, August 28, 2019, https://www.eia.gov/energyexplained/?page=us_energy_home.

19. E2 and Clean Jobs Count, *Clean Jobs America*.

20. Jennifer Liu, "The 10 Fastest-Growing Jobs of the Next Decade—and What They Pay," *Make It*, CNBC.com, September 7, 2019, https://www.cnbc.com/2019/09/07/these-are-the-10-fastest-growing-jobs-of-the-next-decade.html.

21. Dana Varinsky, "Solar-Energy Jobs Are Growing 12 Times as Fast as the US Economy," *Business Insider*, January 26, 2017, https://www.businessinsider.com/solar-energy-job-growth-2017-1.

22. *U.S. Energy and Employment Report* (Washington, DC: US Department of Energy, January 2017), 28, https://www.energy.gov/sites/prod/files/2017/01/f34/2017%20US%20Energy%20and%20Jobs%20Report_0.pdf.

23. Niall McCarthy, "Solar Employs More People in U.S. Electricity Generation Than Oil, Coal and Gas Combined," *Forbes*, January 25, 2017, https://www.forbes.com/sites/niallmccarthy/2017/01/25/u-s-solar-energy-employs-more-people-than-oil-coal-and-gas-combined-infographic/#cb4bd8c28000.

24. Cara Marcy, "Combined Wind and Solar Made Up at Least 20% of Electric Generation in 10 States in 2017," U.S. Energy Information Administration, October 11, 2018, https://www.eia.gov/todayinenergy/detail.php?id=37233.

25. "About Van Jones," Van Jones, https://www.vanjones.net/about/; Dream Corps: Green for All, https://www.thedreamcorps.org/our-programs/green-for-all.

26. Van Jones, "Green for All," February 24, 2010, video, 2:39, https://www.youtube.com/watch?v=RALHsjVQx90.

27. Ryan Dexter (Green for All employee, 2008–2014), email message to author, July 9, 2019.

28. Dexter, email message to author.

29. Ryan Grim, "Glenn Beck Gets First Scalp: Van Jones Resigns," *HuffPost*, May 25, 2011, https://www.huffpost.com/entry/glenn-beck-gets-first-sca_n_278281.

30. Ronald Fel Jones, "Can Solar Shine in Coal Country?" American Solar Energy Society, March 9, 2017, https://www.ases.org/can-solar-shine-in-coal-country.

31. Diane Cardwell, "What's Up in Coal Country: Alternative-Energy Jobs," *New York Times*, September 30, 2017, https://www.nytimes.com/2017/09/30/business/energy-environment/coal-alternative-energy-jobs.html?login=email&auth=login-email.

32. Cardwell, "What's Up in Coal Country."

33. Jeffrey J. Cook et al., *Up to the Challenge: Communities Deploy Solar in Underserved Markets*, National Renewable Energy Laboratory, International City/County Management Association, May 2019, https://www.nrel.gov/docs/fy19osti/72575.pdf.

34. "City Facts: City of Richmond, California," City of Richmond, California, June 14, 2017, http://www.ci.richmond.ca.us/DocumentCenter/Home/View/8348; Maya Raiford Cohen, "Big Oil, Small Town: The Fight for Environmental Justice in Richmond, California," unpublished paper made available to author, December 2017.

35. Allison Cohen et al., "Our Environment, Our Health: A Community-Based Participatory Environmental Health Survey in Richmond, California," *SAGE Journal* 39, no. 2 (2012): 198–209; Cohen, "Big Oil, Small Town."

36. Cohen et al., "Our Environment, Our Health," 198–209.

37. Malia Wollan, "Richmond and Chevron Choose Fork in the Road," *New York Times*, November 1, 2009, https://www.nytimes.com/2009/11/01/us/01sfchevron.html; Cohen, "Big Oil, Small Town."

38. Shalini Kantayya, "Meet the Director of a Solar Documentary You'll Actually Want to Watch," interview by Clayton Aldern, *Grist*, April 1, 2016, https://grist.org/climate-energy/meet-the-director-of-a-solar-documentary-youll-actually-want-to-watch.

39. Kantayya, "Meet the Director of a Solar Documentary."

40. GRID Alternatives, "Our Impact," https://gridalternatives.org/support-us/our-impact, accessed May 1, 2020.

41. GRID Alternatives, "Workforce Development," https://gridalternatives.org/what-we-do/workforce-development, accessed May 1, 2020.

42. Seth Fiegerman, "Tesla Shareholders Approve Solarcity Merger," CNN.com, November 17, 2016, https://money.cnn.com/2016/11/17/technology/tesla-solarcity-merger/index.html.

43. H. J. Mai, "'Opportunities Everywhere': NREL Study Shows Mass Potential for Storage to Provide Peaking Capacity," *Utility Dive*, July 10, 2019, https://www.utilitydive.com/news/opportunities-everywhere-nrel-study-shows-mass-potential-for-storage-to/558344.

44. Tesla, "Tesla Factory," https://www.tesla.com/factory, accessed May 1, 2020.

45. Danielle Muoio, "New Aerial Photos Appear to Show Just How Massive Tesla's Gigafactory Is," *Business Insider*, July 10, 2017, https://www.businessinsider.com/tesla-gigafactory-massive-photos-2017-7.

46. Charles Fleming, "Tesla Gigafactory Worth $100 Billion to Nevada, Governor Says," *Los Angeles Times*, September 4, 2014, https://www.latimes

.com/business/autos/la-fi-hy-tesla-gigafactory-worth-100-billion-to-nevada-governor-says-20140904-story.html.

47. Steve Hanley, "Tesla Now Has 1,800 Employees In New York, Panasonic Quits Gigafactory 2 In Buffalo (The Solar One)," *CleanTechnica*, February 28, 2020, https://cleantechnica.com/2020/02/28/tesla-now-has-1800-employees-in-new-york-panasonic-quits-gigafactory-2-in-buffalo-the-solar-one.

48. Emma Foehringer Merchant, "Progress at Tesla's Gigafactory 2 Remains Murky Amid Concerns over Jobs Target," *Greentech Media*, April 18, 2019, https://www.greentechmedia.com/articles/read/progress-at-teslas-gigafactory-2-murky-amid-concerns-over-jobs-target#gs.phrwrm.

49. Merchant, "Progress at Tesla's Gigafactory 2 Remains Murky Amid Concerns over Jobs Target."

50. Niraj Chokshi and Peter Eavis, "Tesla's Stock Is Up 36% in Two Days. What's Going On?," *New York Times*, February 4, 2020, https://www.nytimes.com/2020/02/04/business/tesla-stock-price.html.

51. UPI, "The Republic Steel Corp., Which Mothballed Steel-Making Operations," UPI Archives, January 17, 1984. https://www.upi.com/Archives/1984/01/17/The-Republic-Steel-Corp-which-mothballed-steel-making-operations-at/2655443163600.

52. Austin Carr and Brian Eckhouse, "Did Elon Musk Forget about Buffalo?," *Bloomberg Businessweek*, November 20, 2018, https://www.bloomberg.com/news/features/2018-11-20/inside-elon-musk-s-forgotten-gigafactory-2-in-buffalo.

53. Chelsea Gohd, "Elon Musk: 100 Tesla Gigafactories Could Power the Entire World," *Futurism*, April 16, 2017, https://futurism.com/elon-musk-100-tesla-gigafactories-could-power-entire-world.

54. GM Corporate Newsroom, "Detroit-Hamtramck to be GM's First Assembly Plant 100 Percent Devoted to Electric Vehicles," January 27, 2020, https://media.gm.com/media/us/en/gm/home.detail.html/content/Pages/news/us/en/2020/jan/0127-dham.html.

55. Jamie L. LaReau, "Here's When You Can Preorder, and See, GMC's New Hummer Electric Pickup," *USA Today*, March 4, 2020, https://www.usatoday.com/story/money/cars/2020/03/04/gmc-electric-hummer-ev-pickup-debut-preorders/4951194002.

56. Andrew J. Hawkins, "GM Is Building an EV Battery Factory with LG Chem in Lordstown, Ohio," *Verge*, December 5, 2019, https://www.theverge.com/2019/12/5/20996866/gm-lg-ev-electric-vehicle-battery-joint-venture-chem-lordstown.

CHAPTER 6: FAITH COMMUNITIES IN ACTION

1. Ellen Levine, *Rachel Carson: A Twentieth-Century Life* (New York: Penguin, 2008), 143.

2. Harold Wood, ed., "Quotations from John Muir," John Muir Exhibit, Sierra Club, 2020, https://vault.sierraclub.org/john_muir_exhibit/writings/favorite_quotations.aspx.

3. "Frequently Asked Questions about the Papal Encyclical," Yale School of Forestry & Environmental Studies. , https://environment.yale.edu/news/article/frequently-asked
-questions-about-the-papal-encyclical, accessed May 1, 2020.

4. Pope Francis, *Laudato Si': On Care for Our Common Home*, Vatican Press, May 24, 2015, verse 217, 158–59, http://w2.vatican.va/content/dam/francesco/pdf/encyclicals/documents/papa-francesco_20150524_enciclica-laudato-si_en.pdf.

5. *Pope Francis, A Man of His Word*, dir. Wim Wenders (Los Angeles: Focus Features, 2018).

6. Pope Francis, *Laudato Si': On Care for Our Common Home*, verse 139, 104.

7. Pope Francis, *Laudato Si': On Care for Our Common Home*, verse 11, 11.

8. Rev. Canon Sally G. Bingham, phone interview with author, December 20, 2019.

9. Rev. Canon Sally G. Bingham, ed., *Love God, Heal Earth* (Pittsburgh: St. Lynn's Press, 2009), vii.

10. Bingham, *Love God, Heal Earth*, v.

11. Bingham, *Love God, Heal Earth*, v.

12. Bingham, *Love God, Heal Earth*, vii.

13. Bingham, *Love God, Heal Earth*, iv.

14. Bingham, *Love God, Heal Earth*, x.

15. Rev. Ambrose Carroll, "Climate Justice," webinar, RE-volv Solar Ambassador College Fellowship, April 21, 2017. The information about and quotes from Ambrose Carroll in the remainder of this section come from this webinar, except where otherwise noted.

16. Carroll, "Climate Justice."

17. Rev. Ambrose Carroll, "Ambrose Carroll, Green the Church Campaign," Pollination Project, https://thepollinationproject.org/grants-awarded/ambrose-carroll-green-the-church-campaign.

18. Carroll, "Climate Justice."

19. Nicollette Higgs, "How Green Theology Is Energizing the Black Community to Fight the Climate Crisis," CNN, November 22, 2019, https://www.cnn.com/2019/11/22/us/green-the-church-black-community-fights-climate-change/index.html.

20. Green the Church, "Green the Church—A Green for All Initiative," http://greenthechurch.org/videos, accessed May 1, 2020.

21. Donella H. Meadows et al., *The Limits to Growth: A Report for the Club of Rome's Project on the Predicament of Mankind* (New York: Universe, 1972).

22. Wisconsin Green Muslims, "About Wisconsin Green Muslims," https://wisconsingreenmuslims.org/about, accessed May 1, 2020.

23. Huda Alkaff, "Huda Alkaff on Wisconsin Green Muslims' Quest for Environmental Justice," interview by Anna M. Gade, Forum on Religion and Ecology at Yale, May 15, 2018, http://fore.yale.edu/news/item/huda-alkaff-on-wisconsin-green-muslims-quest-for-environmental-justice.

24. Kari Lydersen and Yana Kunichoff, "For Milwaukee Muslims, Solar Is an Act of Faith and Service," *Energy News Network*, October 30, 2017, https://energynews.us/2017/10/30/midwest/for-milwaukee-muslims-solar-is-an-act-of-faith-and-service.

25. Wisconsin Green Muslims, "Faith & Solar," https://wisconsingreenmuslims.org/faithsolar, accessed May 1, 2020.

26. White House, "Champions of Change," https://obamawhitehouse.archives.gov/champions/climate-faith-leaders/huda-alkaff, accessed May 1, 2020.

27. *Years of Living Dangerously*, season 1, episode 1, "Dry Season," featuring Harrison Ford and Don Cheadle, prod. James Cameron, Jerry Weintraub, and Arnold Schwarzenegger; aired April 13, 2014, on Showtime. See, also, Katharine Hayhoe, "Years of Living Dangerously Season 1: Bonus Footage—The Crossroads of Climate and Faith," interview by Don Cheadle, video 5:36, https://www.youtube.com/watch?v=nY1HweENTeU.

28. Katharine Hayhoe, "The Most Important Thing You Can Do to Fight Climate Change: Talk about It," TEDWomen, November 2018, video, 17:12, https://www.ted.com/talks/katharine_hayhoe_the_most_important_thing_you_can_do_to_fight_climate_change_talk_about_it?language=en.

29. Hayhoe, "The Most Important Thing You Can Do to Fight Climate Change."

30. Katharine Hayhoe, "Bio," Katharine Hayhoe: Climate Scientist, http://katharinehayhoe.com/wp2016/biography.

31. Katharine Hayhoe and Andrew Farley, *A Climate for Change: Global Warming Facts for Faith-Based Decisions* (New York: Faith Words, 2009), xiii.

32. Hayhoe, "The Most Important Thing You Can Do to Fight Climate Change."

33. Evangelical Environmental Network, "What We Do," Pro-Life Clean Energy Campaign, 2020, https://www.creationcare.org/pro_life_clean_energy_campaign, accessed May 1, 2020.

34. Evangelical Environmental Network, "What We Do."

CHAPTER 7: ENERGY INDEPENDENCE

1. Van Jones, *The Green Collar Economy: How One Solution Can Fix Our Two Biggest Problems* (New York: Harper Collins, 2009), 9.

2. Sierra Hicks, *Powering the Department of Defense: Initiatives to Increase Resiliency and Energy Security* (Washington, DC: American Security Project, September 2017), https://www.americansecurityproject.org/wp-content/uploads/2017/09/Ref-0204-Powering-the-DoD.pdf.

3. Steve Levine, "Isn't Reducing Soldier Casualties a Military Priority?," *Foreign Policy*, January 26, 2011, https://foreignpolicy.com/2011/01/26/isnt-reducing-soldier-casualties-a-military-priority.

4. Steve Hargreaves, "Ambushes Prompt Military to Cut Energy Use," CNN Money, August 16, 2011, https://money.cnn.com/2011/06/14/news/economy/military_energy_strategy/index.htm.

5. Levine, "Isn't Reducing Soldier Casualties a Military Priority?"

6. Hargreaves, "Ambushes Prompt Military to Cut Energy Use."

7. Timothy Gardner, "U.S. Military Marches forward on Green Energy, Despite Trump," Reuters, February 28, 2017, https://www.reuters.com/article/us-usa-military-green-energy-insight/u-s-military-marches-forward-on-green-energy-despite-trump-idUSKBN1683BL.

8. Ryan Koronowski, "Why The U.S. Military Is Pursuing Energy Efficiency, Renewables and Net-Zero Energy Initiatives," *ThinkProgress*, April 4, 2013, https://thinkprogress.org/why-the-u-s-military-is-pursuing-energy-efficiency-renewables-and-net-zero-energy-initiatives-177a32d53323.

9. Lisa A. Jung, *Annual Energy Management and Resilience Report: Fiscal Year 2016*, Office of the Assistant Secretary of Defense for Energy, Installations, and Environment (Washington, DC: Department of Defense, July 2017), 5, https://www.acq.osd.mil/eie/Downloads/IE/FY%202016%20AEMR.pdf.

10. Saltanat Berdikeeva, "The US Military: Winning the Renewable War," *Energy Digital*, September 13, 2017, https://www.energydigital.com/renewable-energy/us-military-winning-renewable-war.

11. Hicks, *Powering the Department of Defense Initiatives*, 1.

12. *Report on Effects of a Changing Climate to the Department of Defense*, Office of the Under Secretary of Defense for Acquisition and Sustainment (Washington, DC: Department of Defense, January 2019), https://climateandsecurity.files.wordpress.com/2019/01/sec_335_ndaa-report_effects_of_a_changing_climate_to_dod.pdf.

13. *Report on Effects of a Changing Climate*, Office of the Under Secretary of Defense, 16.

14. *Report on Effects of a Changing Climate*, Office of the Under Secretary of Defense, 5.

15. *National Security Implications of Climate-Related Risks and a Changing Climate* (Washington, DC: Department of Defense, July 23, 2015), http://archive.defense.gov/pubs/150724-congressional-report-on-national-implications-of-climate-change.pdf?source=govdelivery.

16. RE100, "221 RE100 Companies Have Made a Commitment to Go '100% Renewable,'" http://there100.org/companies.

17. Brad Smith, "Microsoft Will Be Carbon Negative by 2030," Microsoft, https://blogs.microsoft.com/blog/2020/01/16/microsoft-will-be-carbon-negative-by-2030.

18. Matthew Campelli, "New York Yankees Appoint Sport's First Environmental Science Advisor," *Sport Sustainability Journal*, January 30, 2019, https://sportsustainabilityjournal.com/news/new-york-yankees-appoint-sports-first-environmental-science-advisor.

19. Campelli, "New York Yankees Appoint Sport's First Environmental Science Advisor."

20. New York Yankees (@Yankees), "Today, the Yankees became the 1st major North American sports team to sign on to the UN Sports for Climate

Action Framework," Twitter, April 3, 2019, 9:51 a.m., https://twitter.com/Yankees/status/1113453954405416962.

21. New York Yankees, "Sustainability Initiatives at Yankee Stadium," Yankees.com, https://www.mlb.com/yankees/ballpark/information/sustainability-initiatives, accessed March 8, 2020.

22. Dan Hernandez, "Las Vegas Casinos Seek to Power Their Bright Lights with Renewable Energy," *Guardian*, March 7, 2016, https://www.theguardian.com/environment/2016/mar/07/las-vegas-casinos-solar-power-nevada-energy.

23. Sam Pothecary, "Largest Solar Rooftop System in the US Installed on the Mandalay Bay Convention Center," *PV Magazine*, July 7, 2016, https://pv-magazine-usa.com/2016/07/07/largest-solar-rooftop-system-in-the-us-installed-on-the-mandalay-bay-convention-center.

24. Hernandez, "Las Vegas Casinos Seek to Power Their Bright Lights with Renewable Energy."

25. Pothecary, "Largest Solar Rooftop System in the US."

26. Hernandez, "Las Vegas Casinos Seek to Power Their Bright Lights with Renewable Energy."

27. Mike O'Boyle, "Three Ways Electric Utilities Can Avoid a Death Spiral," *Forbes*, September 25, 2017, https://www.forbes.com/sites/energyinnovation/2017/09/25/three-ways-electric-utilities-can-avoid-a-death-spiral/#34976b2f758d.

28. Hernandez, "Las Vegas Casinos Seek to Power Their Bright Lights with Renewable Energy."

29. Sean Whaley, "MGM Resorts to Leave Nevada Power, Pay $86.9M Exit Fee," *Las Vegas Review-Journal*, May 19, 2016, https://www.reviewjournal.com/business/energy/mgm-resorts-to-leave-nevada-power-pay-86-9m-exit-fee.

30. Christian Roselund, "100 MW MGM-Invenergy Solar Project to Power Las Vegas Strip," *PV Magazine*, April 20, 2018, https://pv-magazine-usa.com/2018/04/20/100-mw-mgm-invenergy-solar-project-to-power-las-vegas-strip.

31. Hernandez, "Las Vegas Casinos Seek to Power Their Bright Lights with Renewable Energy."

32. US Energy Information Administration, "Hawaii State Profile and Energy Estimates," December 19, 2019, https://www.eia.gov/state/analysis.php?sid=HI.

33. Megan Fernandes, "Study: Hawaii Has the Highest Electricity Prices in the Country," *Pacific Business News*, July 3, 2019, https://www.bizjournals.com/pacific/news/2019/07/03/study-hawaii-has-the-highest-elec.html.

34. Marco Mangelsdorf, "The New Normal for Rooftop Solar in Hawaii?," *Greentech Media*, July 14, 2017, https://www.greentechmedia.com/articles/read/the-new-normal-for-rooftop-solar-in-hawaii#gs.3zvy5f.

35. Mangelsdorf, "The New Normal?"

36. Julian Spector, "How Energy Storage's Growth Trajectory Differs from the Early Days of Solar," *Greentech Media*, December 4, 2019, https://www.greentechmedia.com/articles/read/energy-storage-summit-growth-trajectory.

37. Cole Mellino, "Hawaii Enacts Nation's First 100% Renewable Energy Standard," EcoWatch, June 11, 2015, https://www.ecowatch.com/hawaii-enacts-nations-first-100-renewable-energy-standard-1882047718.html.

38. Iulia Gheorghiu, "Puerto Rico Passes 100% Renewable Energy Bill as It Aims for Storm Resilience," *Utility Dive*, March 26, 2019, https://www.utilitydive.com/news/puerto-rico-passes-100-renewable-energy-bill-as-it-aims-for-storm-resilien/551303.

39. Jeremy Rifkin, *Third Industrial Revolution: How Lateral Power Is Transforming Energy, the Economy, and the World*, (New York: St. Martin's Press, 2011), 115.

40. David Biello, "Behind the Light Switch: What Will a Smart Grid Look Like?" *Scientific American*, May 12, 2010, https://www.scientificamerican.com/article/what-will-smart-grid-look-like.

41. Gheorghiu, "Puerto Rico Passes 100% Renewable Energy Bill."

42. Iulia Gheorghiu, "Puerto Rico Proposes Largest Solar, Storage Buildout in US with 20-Year Draft Resource Plan," *Utility Dive*, February 6, 2019, https://www.utilitydive.com/news/puerto-rico-proposes-largest-solar-storage-buildout-in-us-with-20-year-dra/547781.

43. Emma Foehringer Merchant, "Final IRP Proposal for Puerto Rico Calls for 'Mini-Grids' and Rapid Solar and Storage Deployment," *Greentech Media*, February 15, 2019, https://www.greentechmedia.com/articles/read/final-irp-proposal-for-puerto-rico-calls-for-mini-grids-and-rapid-solar-and#gs.408ovv.44. Gheorghiu, "Puerto Rico Proposes Largest Solar."

44. "One Billion People Don't Have Access to Electricity and This Map Shows You Who," *Mashable*, September 2017, https://mashable.com/2017/09/15/one-billion-people-dont-have-access-to-electricity.

45. Mike Murphy, "Cellphones Now Outnumber the World's Population," *Quartz*, April 29, 2019, https://qz.com/1608103/there-are-now-more-cellphones-than-people-in-the-world.

46. Michael Bloomberg and Carl Pope, *Climate of Hope: How Cities, Businesses, and Citizens can Save the Planet* (New York: St. Martin's Press, 2017), 199.

CHAPTER 8: COMMUNITIES AT THE FOREFRONT

1. Damian Carrington, "'Our Leaders Are Like Children,' School Strike Founder Tells Climate Summit," *Guardian*, December 4, 2018, https://www.theguardian.com/environment/2018/dec/04/leaders-like-children-school-strike-founder-greta-thunberg-tells-un-climate-summit.

2. Karl Evers-Hillstrom, "Lobbying Spending Reaches $3.4 Billion in 2018, Highest in 8 Years," Open Secrets, January 25, 2019, https://www.opensecrets.org/news/2019/01/lobbying-spending-reaches-3-4-billion-in-18.

3. Maxine Bédat and Michael Shank, "Every Purchase You Make Is a Chance to Vote with Your Wallet," *Fast Company*, April 5, 2017, https://www.fastcompany.com/40402079/every-purchase-you-make-is-a-chance-to-vote-with-your-wallet.

4. Heath and Heath, *Switch*, 254–55.

5. Carolyn Kormann, "Greening the Tea Party," *New Yorker*, February 17, 2015, https://www.newyorker.com/tech/annals-of-technology/green-tea-party-solar.

6. Wendy Koch, "How Keeping Up with the Joneses Could Save You Money," *National Geographic*, July 17, 2015, https://www.nationalgeographic.com/news/energy/2015/07/150717-using-peer-pressure-to-cut-energy-usage.

7. Aaron Mondry, "Welcome to an American City Where the Government Barely Exists," *Splinter News*, September 22, 2017, https://splinternews.com/welcome-to-an-american-city-where-the-government-barely-1818667220.

8. Mondry, "Welcome to an American City Where the Government Barely Exists."

9. Mondry, "Welcome to an American City Where the Government Barely Exists."

10. Mondry, "Welcome to an American City Where the Government Barely Exists."

11. Jackson Koeppel, "Organizing for Energy Democracy in the Face of Austerity," Common Dreams, April 14, 2019, https://www.commondreams.org/views/2019/04/14/organizing-energy-democracy-face-austerity; Soulardarity, "Weatherization," https://www.soulardarity.com/weatherization.

12. Neville Williams, *Chasing the Sun: Solar Adventures Around the World* (Gabriola Island, BC: New Society, 2005), 39.

13. GRID Alternatives, "Our Impact," https://gridalternatives.org/support-us/our-impact.

14. GRID Alternatives, "Mission and History," https://gridalternatives.org/who-we-are/mission-history.

15. Majora Carter, "Greening the Ghetto," TED Talk, video, 18:17, https://www.ted.com/talks/majora_carter_greening_the_ghetto/transcript#t-167642.

16. Emma Foehringer Merchant, "Report Finds Wide Racial and Ethnic Disparities in Rooftop Solar Installations," *Greentech Media*, January 14, 2019, https://www.greentechmedia.com/articles/read/report-finds-wide-racial-and-ethnic-disparities-in-rooftop-solar#gs.iabv45.

17. Solar Foundation, "Diversity and Inclusion in the U.S. Solar Industry," infographic, 2019, https://www.thesolarfoundation.org/wp-content/uploads/2019/05/Solar-Diversity-Infographic.pdf.

18. Merchant, "Report Finds Wide Racial and Ethnic Disparities in Rooftop Solar Installations."

19. Merchant, "Report Finds Wide Racial and Ethnic Disparities in Rooftop Solar Installations."

20. Merchant, "Report Finds Wide Racial and Ethnic Disparities in Rooftop Solar Installations."

21. NAACP, "Go Solar in Your Community," https://www.naacp.org/go-solar-pledge.

22. Solar United Neighbors, "Detailed History: Our Story," https://www.solarunitedneighbors.org/about-us/our-history/detailed-history.

23. Solar United Neighbors, "Our Impact," https://www.solarunitedneighbors.org/about-us/our-impact, accessed February 13, 2020.

24. Solutions Project, "How We Turned an Impossible Idea into a Global Movement," https://thesolutionsproject.org/our-story, accessed March 8, 2020.

25. Mark Ruffalo, "Why I Fight Against Fracking," Earthjustice, April 15, 2011, https://earthjustice.org/blog/2011-april/mark-ruffalo-why-i-fight-against-fracking.

26. Solutions Project, "How We Turned an Impossible Idea into a Global Movement."

27. Mark Z. Jacobson et al., "100% Clean and Renewable Wind, Water, and Sunlight (WWS) All-Sector Energy Roadmaps for the 50 United States," abstract, *Energy & Environmental Science* 7 (2015), https://pubs.rsc.org/en/content/articlelanding/2015/EE/C5EE01283J#!divAbstract.

28. Mark Z. Jacobson, et al., "100% Clean and Renewable Wind, Water, and Sunlight All-Sector Energy Roadmaps for 139 Countries of the World," *Joule*, September 6, 2017, https://web.stanford.edu/group/efmh/jacobson/Articles/I/CountriesWWS.pdf.

29. Brian Willis, "With the Retirement of Dolet Hills, Sierra Club's Beyond Coal Campaign Retires 300th Coal Plant Since 2010," Sierra Club, January 8, 2020, https://www.sierraclub.org/press-releases/2020/01/retirement-dolet-hills-sierra-club-s-beyond-coal-campaign-retires-300th-coal?_ga=2.3273136.1011338144.1578771807-1527504607.1578771807.

30. Sierra Club, "100% Campaign," Ready for 100, https://www.sierraclub.org/ready-for-100/campaign.

31. Sierra Club, "100% Campaign."

32. *Progress Toward 100% Clean Energy in Cities and States Across the US* (Los Angeles: UCLA Luskin Center for Innovation, November 2019), https://innovation.luskin.ucla.edu/wp-content/uploads/2019/11/100-Clean-Energy-Progress-Report-UCLA-2.pdf.

33. Xcel Energy, "Who We Are," https://www.xcelenergy.com/company/corporate_responsibility_report/who_we_are, accessed March 8, 2020.

34. Kelsey Misbrener, "APS Plans to Retire Coal by 2031, Reach 100% Clean Energy by 2050," *Solar Power World*, January 22, 2020, https://www.solarpowerworldonline.com/2020/01/aps-sets-goal-100-percent-clean-energy-by-2050.

35. Julia Pyper, "A Conversation with David Crane on Getting Fired from NRG and What's Next for His Energy Plans," *Greentech Media*, April 29, 2016, https://www.greentechmedia.com/articles/read/A-Conversation-with-David-Crane.

36. Pyper, "A Conversation with David Crane."

37. Brené Brown, *Daring Greatly: How the Courage to Be Vulnerable Transforms the Way We Live, Love, Parent, and Lead* (New York: Avery, 2012), 1.

38. Andrew Birch, "How to Halve the Cost of Residential Solar in the US," *Greentech Media*, January 5, 2018, https://www.greentechmedia.com/articles/read/how-to-halve-the-cost-of-residential-solar-in-the-us#gs.j7xg88.

39. Thad Culley, "South Carolina Senate Passes Energy Freedom Act," Vote Solar, May 8, 2019, https://votesolar.org/usa/south-carolina/updates/south-carolina-senate-passes-energy-freedom-act.

40. MCE, "Your LOCAL Energy Choice," https://www.mcecleanenergy.org/about-us.

41. Rebecca Bowe and Dan Brekke, "Behind Props. G and H, Dueling S.F. 'Green' Energy Ballot Measures," KQED, Nov 2, 2015, https://www.kqed.org/news/10590903/citys-cleanpowersf-program-central-to-upcoming-ballot-battle.

42. John Farrell, "Why Minnesota's Community Solar Program Is the Best," Institute for Local Self-Reliance, February 24, 2020, https://ilsr.org/minnesotas-community-solar-program.

43. Michael T. Deane, "Top 6 Reasons New Businesses Fail," Investopedia, June 25, 2019, https://www.investopedia.com/slide-show/top-6-reasons-new-businesses-fail.

44. Elon Musk, "Eating Glass and Starting Up," January 28, 2015, YouTube video, 2:15, https://www.youtube.com/watch?v=yZlHbjxtECg.

45. Danny Kennedy, "CEC Approves $2.7 Million to Help Entrepreneurs Bring Clean Energy Technologies Closer to Market," California Energy Commission, January 22, 2020, https://www.energy.ca.gov/news/2020-01/cec-approves-27-million-help-entrepreneurs-bring-clean-energy-technologies.

46. PV Magazine, "#Solar100's Emily Kirsch: The Oprah of Clean Energy," January 23, 2019, https://pv-magazine-usa.com/2019/01/23/solar100s-emily-kirsch-the-oprah-of-clean-energy.

47. Danny Kennedy, *The Rooftop Revolution: How Solar Power Can Save Our Economy—and Our Planet—from Dirty Energy* (San Francisco: Berrett-Koehler, 2012), 58–59.

48. Alan Thein Durning, "Are We Happy Yet?," in *Ecopsychology: Restoring the Earth, Healing the Mind*, ed. Theodore Roszak et al., Sierra Club Books (San Francisco: Counterpoint, 1995), 76.

CHAPTER 9: GRATITUDE, SIMPLICITY, AND SERVICE

1. David R. Loy and John Stanley, *A Buddhist Response to the Climate Emergency* (Somerville, MA: Wisdom Publications, 2009), 9.

2. James Gustave Speth, *The Bridge at the Edge of the World: Capitalism, the Environment, and Crossing from Crisis to Sustainability* (New Haven, CT: Yale University Press, 2008), 65.

3. Kevin Kelleher, "Gilded Age 2.0: U.S. Income Inequality Increases to Pre-Great Depression Levels," *Fortune*, February 13, 2019, https://fortune.com/2019/02/13/us-income-inequality-bad-great-depression.

4. Noah Kirsch, "The 3 Richest Americans Hold More Wealth Than Bottom 50% of the Country, Study Finds," *Forbes*, November 9, 2017, https://www.forbes.com/sites/noahkirsch/2017/11/09/the-3-richest-americans-hold-more-wealth-than-bottom-50-of-country-study-finds/#c60916a3cf86.

5. Olga Khazan, "The True Cause of the Opioid Epidemic," *Atlantic*, January 2, 2020, https://www.theatlantic.com/health/archive/2020/01/what-caused-opioid-epidemic/604330.

6. Glenn Sullivan, "Thoughts on the Suicide Epidemic," *Psychology Today*, March 31, 2019, https://www.psychologytoday.com/us/blog/acquainted-the-night/201903/thoughts-the-suicide-epidemic.

7. Felix Gussone, "America's Obesity Epidemic Reaches Record High, New Report Says," NBC News, October 12, 2017, https://www.nbcnews.com/health/health-news/america-s-obesity-epidemic-reaches-record-high-new-report-says-n810231.

8. Briony Harris, "These Are the Happiest Countries in the World," World Economic Forum, March 16, 2018, https://www.weforum.org/agenda/2018/03/these-are-the-happiest-countries-in-the-world. Access the full 2018 *World Happiness Report* online at https://s3.amazonaws.com/happiness-report/2018/WHR_web.pdf.

9. Harris, "These Are the Happiest Countries."

10. Harris, "These Are the Happiest Countries."

11. "Time Flies: U.S. Adults Now Spend Nearly Half a Day Interacting with Media," Nielsen, July, 31, 2018, https://www.nielsen.com/us/en/insights/article/2018/time-flies-us-adults-now-spend-nearly-half-a-day-interacting-with-media.

12. Caitlin Johnson, "Cutting Through Advertising Clutter," *CBS Sunday Morning*, September 17, 2006, https://www.cbsnews.com/news/cutting-through-advertising-clutter.

13. University of Chicago Press Journals, "How Do Beauty Product Ads Affect Consumer Self Esteem and Purchasing?," *ScienceDaily*, October 26, 2010, https://www.sciencedaily.com/releases/2010/10/101018163112.htm.

14. Alan Thein Durning, *How Much Is Enough? The Consumer Society and the Future of the Earth*, Worldwatch Environmental Alert Series (New York: Norton, 1992), 23.

15. Mihaly Csikszentmihalyi, *Flow: The Psychology of Optimal Experience* (New York: Harper Perennial, 1990), 2.

16. Csikszentmihalyi, *Flow*, 23.

17. Paul Hawken, *Drawdown: The Most Comprehensive Plan Ever Proposed to Reverse Global Warming* (New York: Penguin, 2017), 42.

18. James Randerson, "The Path to Happiness: It Is Better to Give Than Receive," *Guardian*, March 20, 2008, https://www.theguardian.com/science/2008/mar/21/medicalresearch.usa.

19. Shawn Achor, *The Happiness Advantage: How a Positive Brain Fuels Success in Work and Life* (New York: Currency, 2010), 52, 201.

20. Ian Sample, "Our Flexible Friend," *Guardian*, April 6, 2009, https://www.theguardian.com/science/2009/apr/07/brain-neuroscience-stroke-depression.

21. Achor, *The Happiness Advantage*, 89.

22. Achor, *The Happiness Advantage*, 98.

23. Achor, *The Happiness Advantage*, 101.

24. Elizabeth Witherell, ed., "The Writings of Henry D. Thoreau," Thoreau Edition Project, University of California–Santa Barbara, http://thoreau.library.ucsb.edu/thoreau_main.html.

25. Duane Elgin, *Voluntary Simplicity: Toward a Way of Life That Is Outwardly Simple, Inwardly Rich* (New York: Quill, 1993), 45.

26. Marie Kondo, *The Life-Changing Magic of Tidying Up: The Japanese Art of Decluttering and Organizing* (Berkeley, CA: Ten Speed Press, 2014), 41.

27. Kondo, *The Life-Changing Magic of Tidying Up*, 42.

28. Kondo, *The Life-Changing Magic of Tidying Up*, 4.

29. Kondo, *The Life-Changing Magic of Tidying Up*, 3.

30. Mark Abadi, "6 American Work Habits People in Other Countries Think Are Ridiculous," *Business Insider*, November 17, 2017, https://www.independent.co.uk/news/business/american-work-habits-us-countries-job-styles-hours-hoilday-a8060616.html.

31. Alan Watts, "What If Money Didn't Matter?," Neovich, YouTube, audio, 3:10, https://www.youtube.com/watch?v=Ddf5oWswq6E.

32. Ilya Pozin, "The Secret to Happiness? Spend Money on Experiences, Not Things," *Forbes*, March 3, 2016, https://www.forbes.com/sites/ilyapozin/2016/03/03/the-secret-to-happiness-spend-money-on-experiences-not-things/#5e95be3f39a6.

33. Diane Dreher, "Why Nature Is Good for Our Brains," *Psychology Today*, November 10, 2015, https://www.psychologytoday.com/us/blog/your-personal-renaissance/201511/why-nature-is-good-our-brains.

34. Alice G. Walton, "7 Ways Meditation Can Actually Change the Brain," *Forbes*, February 9, 2015, https://www.forbes.com/sites/alicegwalton/2015/02/09/7-ways-meditation-can-actually-change-the-brain/#2127b39f1465.

35. Csikszentmihalyi, *Flow*, 4.

36. Csikszentmihalyi, *Flow*, 52.

37. Frederick Douglass, "Treatment of Slaves on Lloyd's Plantation," chapter 6 in *My Bondage and My Freedom*, Literature Network, http://www.online-literature.com/frederick_douglass/bondage_freedom/6.

38. Yuri Kochiyama, Bus Stop Billboard 2019, a project of the San Francisco Arts Commission.

39. Achor, *The Happiness Advantage*, 52.

40. Angela Duckworth, *Grit: The Power of Passion and Perseverance* (New York: Scribner, 2016), 147.

41. Duckworth, *Grit*, 144, 149.

42. Martin Luther King Jr., *I Have a Dream: Writings and Speeches That Changed the World* (New York: HarperCollins, 1986), 182.

43. King, *I Have a Dream*, 184–85.

44. King, *I Have a Dream*, 189.

45. John Muir, *My First Summer in the Sierra* (Cambridge, MA: Riverside Press, 1911), 110.

46. Martin Luther King Jr., "Letter from a Birmingham Jail," Martin Luther King Jr. Research and Education Institute, Stanford University, https://kinginstitute.stanford.edu/king-papers/documents/letter-birmingham-jail.

47. Mahatma Gandhi, *The Essential Gandhi, An Anthology of His Writings on His Life, Work, and Ideas* (New York: Vintage Spiritual Classics, 2002), 64.

48. King, *I Have a Dream*, 148.

CHAPTER 10: ROLLING UP OUR SLEEVES

1. Dalai Lama XIV, *A Buddhist Response to the Climate Emergency* (Somerville, MA: Wisdom Publications, 2009), 22.

2. King, *Strength to Love*, 51.

3. Jesse Pound, "Cramer Sees Oil Stocks in the 'Death Knell Phase,' Says They Are the New Tobacco," CNBC, January 31, 2020, https://www.cnbc.com/2020/01/31/cramer-sees-oil-stocks-in-the-death-knell-phase-says-new-tobacco.html.

4. Bill McKibben, "'A Bomb in the Center of the Climate Movement': Michael Moore Damages Our Most Important Goal," *Rolling Stone*, May 1, 2020, https://www.rollingstone.com/politics/political-commentary/bill-mckibben-climate-movement-michael-moore-993073.

5. Michael Braungart and William McDonough, *Cradle to Cradle: Remaking the Way We Make Things* (New York: North Point Press, 2002).

6. Michael Pollan, *Food Rules: An Eater's Manual* (New York: Penguin, 2009).

7. "How to Eat: Diet Secrets from Michael Pollan," review of *Food Rules*, *Houston Chronicle*, January 23, 2010, https://michaelpollan.com/reviews/how-to-eat.

8. Hawken, *Drawdown*, 39, 40.

9. "Key Facts and Findings," Food and Agriculture Organization of the United Nations, http://www.fao.org/news/story/en/item/197623/icode, accessed May 21, 2020; Damian Carrington, "Avoiding Meat and Dairy Is 'Single Biggest Way' to Reduce Your Impact on Earth," *Guardian*, May 31, 2018, https://www.theguardian.com/environment/2018/may/31/avoiding-meat-and-dairy-is-single-biggest-way-to-reduce-your-impact-on-earth.

10. Umair Irfan, "Report: We Have to Change How We Eat and Grow Food to Fight Climate Change," *Vox*, August 8, 2019, https://www.vox.com/2019/8/8/20758461/climate-change-report-2019-un-ipcc-land-food.

11. Hawken, *Drawdown*, 42.

12. Robert D. Putnam, *Bowling Alone: The Collapse and Revival of American Community* (New York: Simon & Schuster, 1999), 268.

13. "Victory Gardens in World War II," Kraus and Speers Cornucopia of History, February 13, 2014, video, 1:37, https://youtu.be/Lnwle4dsEgc.

14. Joanna Macy, *World as Lover, World as Self: Courage for Global Justice and Ecological Renewal* (Berkeley, CA: Parallax Press, 2007), 76.

INDEX

Note: Numbers in italics refer to tables.